AF545146

EXPLORING MILITARY SERVICE FOR WOMEN

by
Mary McGowan Slappey

THE ROSEN PUBLISHING GROUP
New York

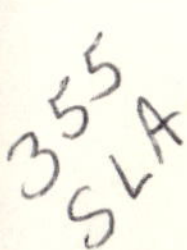

Published in 1986, 1989 by The Rosen Publishing Group, Inc.
29 East 21st Street, New York, NY 10010

Revised Edition 1989

Library of Congress Cataloging-in-Publication Data

Slappey, Mary McGowan, 1914-
Exploring military service for women.

(Exploring careers)
Includes index.
1. United States—Armed Forces—Women.
2. United States—Armed Forces—Vocational guidance.
I. Title II. Series: Exploring careers (Rosen Publishing Group)
UB418.W65S57 355'.0088042 86-3254
ISBN 0-8239-0996-4

Manufactured in the United States of America

To my first cousin, Dr. Williard Slappey, and his wife, Helene Mayo Slappey, of North Carolina...Army nurse and veterinary doctor who became the parents of four sons and have a number of grandchildren.

"I welcome this opportunity to highlight the many contributions women are making to the national security of the United States... I believe it is of vital importance that Americans be aware of the work that women in the uniformed services and civilian force continue to perform for us...The Department of Defense is dedicated to the principle that the individual has infinite dignity and worth. Fair and equitable treatment for women is part of our DoD Human Goals Program, and it is a goal we intend to fulfill completely."

—Caspar W. Weinberger

Secretary of Defense*, 1981-1988

"We can no longer go to war without the women."

—Lt. General Colin Powell

Director, National Security Council

---

*From *Going Strong, Women in Defense*, published by the Department of Defense.

The author wishes to acknowledge the helpful cooperation of the following:

Lieutenant Colonel B. Rothget
U.S. Air Force
Chief, Public Affairs
Pentagon, Washington, D.C.

Lieutenant Colonel F.C. Peck
U.S. Marine Corps
Head, Media Branch
Public Affairs Division

Lieutenant Colonel Benjamin T. Layton
Kensington, Maryland

Andrea Hamburger
Public Affairs Specialist
United States Military Academy
West Point, New York

Joann R. Rachtenbach
Staff Officer
Special Activities Media
Office of the Secretary of the Army
Washington, D.C.

Anna C. Urband
Assistant Head, Media Services Branch
Department of the Navy
Office of Information
Washington, D.C.

TSgt. Timothy A. Thomas
Air Force Reserve Recruiting Office
Alexandria, Va.

Nicholas G. Sandefer
U.S. Coast Guard Headquarters
Washington, D.C.

Constance Manchester Ellison
Literary Consultant
(Mrs. Ellison's late husband was an Army Architect, WWI)

Staff Sergeant Clay F. Miller
Station Commander
U.S. Army Recruiting Station
Bethesda, Md.

Captain G. Lacey
Washington Recruiting Company
Bethesda, Maryland

Jim Sandefur
Deputy Chief, Media Relations
Office of Public Affairs
Randolph Air Force Base, Texas

About the Author

Mary McGowan Slappey is a pioneer in the history of women in the United States Navy. In 1943 she became a member of the first classes of women enrolled and trained as Waves—Women Accepted for Volunteer Emergency Service. After attending Indoctrination School at Mt. Holyoke, Massachusetts, as an apprentice seaman, she received her first Navy rank of Ensign, and her first assignment, to Base personnel, Base Maintenance. On the face of it a prosaic job, what it actually involved was maintenance of bases in the far corners of the world. Within a few weeks she was assigned to administer the worldwide Seabee Construction Engineering program, directly responsible to the Chief of Naval Operations. The Seabees grew, from 90,000 to more than 200,000 the construction men of the Navy. At war's end Lieutenant (j.g.) Slappey received a commendation from the Secretary of the Navy for "outstanding performance of duty. . . . By her initiative and administrative ability, [she] contributed materially to the successful prosecution of the war and upheld the highest traditions of the United States Naval Service."

Retiring with the rank of lieutenant commander in the Naval Reserve, she attended law school and studied fine arts, later working as an editor and teacher. She is listed in *Who's Who Women in the World*, *American Women*, and other biographical directories of authors and journalists. She holds a Cultural Doctorate of Literature from the World University and is a member of the Institute of Achievement of the American Biographical Association, Raleigh, North Carolina.

Contents

Chapter I

You and the Military

Have you ever thought of joining the military? In a few years you will be a high school graduate and ready to make a career choice. There are many opportunities for service women. You could fly with the Air Force, study oceanography with the Navy, march with the Army, serve in support roles in the Marine Corps or Coast Guard. You could be a nurse, a flight attendant, a journalist, a graphics expert, or a bugle player—all in the military. You could repair airplanes or operate computers. You might even choose to go to sea or study foreign service and arrive at NATO.

Sound like a movie script? Not exactly. A service career requires hard work and discipline, but as Kipling wrote to young boys in "If":

> If you can dream—and not make dreams your master
> .... If you can meet with Triumph and Disaster
> And treat those two impostors just the same ...
> If you can fill the unforgiving minute
> With sixty seconds' worth of distance run,
> Yours is the Earth and everything that's in it,
> And—which is more—you'll be a Man, my son!

And we can paraphrase that: "You'll be a woman, my girl."

You may already know something about leadership and teamwork from basketball and school, but the services are prepared to teach you more. Growing up, however, doesn't all have to be arduous. Growing up can be fun as you find your skills and develop your own potential.

Perhaps when you saw the ROTC marching at your school, you thought of wearing a uniform. Perhaps you have a father or an uncle, a brother, a mother, or an aunt who served in the Army or

Navy and you like their stories and their air of confidence. Maybe you entered a speaking or writing contest about American patriotism, our flag, or our Constitution, and won a prize.

Well, if you really want to join one of the services, it doesn't have to be just a dream. It can come true. This book is written to tell you more about it and help you to make wise choices that can lead to more education, adventure, and a lifetime career with security. Some of those choices you need to make now while you're still in high school or junior high.

If you really want something in life, the first step is to learn as much about it as possible. What are some of the reasons for joining one of the services. Let's list them:

1. *Patriotism.* You must love your country and the freedom for which it stands to serve it well. Loyalty to your country and your service is a fundamental requirement if you are to get the most out of your service career. In later chapters we will tell you something of the history of each of the modern-day services that now offer twentieth-century opportunities for women.

2. *Educational and social opportunities.* Currently there are 199,000 women in service, and they are a part of the regular Defense Department as well as the branch in which they serve. A variety of educational and training opportunities has been designed to fit the talents of each woman candidate. We'll tell you more about those later and also introduce you to some of the girls and women who are enjoying their jobs, both traditional and nontraditional. Your service career might even lead you to become a service wife. One I know, Mrs. Dorothea Johnson, has written a book, *Entertaining with Ease and Elegance*, and gives courses to servicewomen and others on being a good hostess.

3. *Promotions, job security, world travel.* Even routine jobs can lead to promotions, and with new skills and learning can come opportunities to be stationed in other parts of the country or even to see the world. Adventure with some security is a hard package to find, but many young women are finding it in military careers.

4. *Achieving your goals with benefits along the way.* Health care, subsistence allowances, the honor and dignity of wearing the uniform of your country, the chance to make friends, to develop poise as you further your own special interests and talents are all positive factors. If you're recruited, you'll want to study the opportunities available so you can help the counselors fit you into the

right billet—the one that will lead you where you want to go: doctor, nurse, engineer, mechanic, foreign service, electronics, computers, music, graphics, communications, journalism, and others.

Once you're sworn in, you are no longer a civilian. As a military trainee, you are subject to military law and regulations. You will be held to a code of ethics and honor. You will learn to simplify your life-style. You will be expected to wear your uniform with honor and dignity and keep it in good condition. You will have to keep your hair off your collar. You will have to follow the uniform of the day.

You will have to get along with others and to keep military secrets. Although personal ambition may have led you here, as a servicewoman you will learn more about teamwork and leadership and the chain of command. You will learn to take orders, and if you learn to listen well you may be promoted to giving orders yourself.

Your first training days are important because they are the beginning of a service career that can lead to higher achievement. But at first you will have to apply yourself with all the energy and enthusiasm you can muster to meet the challenge of the opportunity offered you. There will be marching and vaccinations and fire drills and lessons, but you'll probably get a thrill seeing yourself in uniform. You'll enjoy meeting girls from far and near, and men comrades too, and sharing experiences at leisure times, on chapel days, or at special social events—more frequent after that first basic training.

Even to parade music, it's one step at a time!

The day you're sworn in to your special service and know that you've met the requirements, that you've taken that first big step, your folks will start talking about their daughter in the Service, and you'll be excited as new doors open for you and you are launched on your own career of service.

The Military Oath of Enlistment

> I do solemnly swear that I will support and defend the Constitution of the United States against all enemies, foreign and domestic; that I will bear true faith and allegiance to the same; and that I will obey the orders of the President of the United States and the orders of the officers appointed over me, according to regulations and the Uniform Code of Military Justice. So help me God.

The New Opportunities for Women

One of the great achievements of the military has been to elevate the status of women. In the years since World War II, more progress has been made toward equal rights than was made in all the thousands of years before. It was only a century ago that Susan B. Anthony and like-minded suffragists were active and only sixty-five years ago that we won the right to vote.

It is shocking to realize the low estate that women held legally throughout history. As their husband's possession, their position

U.S. ARMY PHOTO

A signal repairwoman checks the truck telephone switching system at the U.S. Army Communications Command, Ft. Richardson, Alaska.

was not much better than that of a common chattel. But all that has changed. Women are succeeding in higher roles in the military as well as in private life. The recent appointment of Gail M. Reals as a Brigadier General and Base Commander of the Quantico Marine Corps is a fine example.

The goal for every individual to be allowed to function to his or her highest potential is in sight. Biologically and genetically women's capabilities and brain power are equal to any man's. Exclusion from combat duty is occasioned only by differences in physical strength, present unavailability of facilities, and the desire of a democratic country to protect its military service women as much as possible.

In response to the need to further integrate women into the services, Secretary of Defense Frank C. Carlucci in 1988 set up a special study group. Headed by David J. Armor, Deputy Assistant Secretary of Defense, the group included senior officials from each military department, the Joint Chiefs of Staff, the National Security Council, and the Defense Advisory Committee on Women in the Services. Secretary Armor announced: "We're developing a clear rationale for opening all jobs except those that are strictly combat." Among highlights of the proposed reforms, the Air Force will allow women aboard high-altitude reconnaissance aircraft; the Marine Corps will have women serving as security guards at U.S. embassies, and the Navy will permit female personnel on the EP-3 reconnaissance aircraft. The Navy's decision to assign women to ammunition ships, oilers, and other vessels in the "combat logistics force" should mean some 9,000 new jobs for women sailors, and 4,000 other jobs are expected to be opened as a result of the study.

New opportunities will abound, so start your career with confidence. Top defense planners are on your side ready to recognize your skills, your dedication, and your right to demand the respect you deserve and to gain the promotions you earn.

The future is yours. The brass ring is waiting, and the hash marks, maybe even the gold epaulets.

Chapter II

The United States Army

"My fellow Americans, ask not what your country can do for you—ask what you can do for your country."

— President John F. Kennedy

The U.S. Post Office has issued a stamp in honor of Bostonian Henry Knox (1750–1806), one of George Washington's closest friends and advisers. A specialist in artillery, Knox rose to the rank of general during the Revolutionary War and took part in nearly every major engagement. He joined the Boston Grenadier Corps in 1772 and became a student of military science and engineering, volunteering for the Continental Army three years later. It was he who directed Washington's crossing of the Delaware in 1776, commemorated so vividly in painting and prose.

In 1779 General Knox proposed the establishment of the military academy that became West Point. He was the first Secretary of War under the Articles of Confederation and under the Constitution. His idea for a national militia to protect the country became the basis for the National Guard. Fort Knox, the U.S. gold depository, and the city of Knoxville, Tennessee, were named for him.

West Point

West Point was founded on a site that held strategic importance for American troops during the Revolution. General Washington selected the site for the nation's first military academy, and President Jefferson directed the formal opening on July 4, 1802. A beautiful 2,500-acre campus in Orange County, New York, approximately 50 miles from New York City, the cadet area framed by the Hudson highlands overlooks the Hudson River at West Point. The main cadet complex maintains a harmonious blend of Gothic-style architecture and the natural beauty of the surrounding hills, green in spring, orange and gold in autumn, and stark and white in winter.

The focal point of the cadet area is Washington Hall, the dining hall where the entire Corps assembles three times a day for family style (and very hearty) meals. In addition to the dining hall, the building contains two academic departments, classrooms, and cadet support facilities.

Flanking Washington Hall on the north, east, and south is the barracks complex, where cadets live two per room.

Adjacent to the cadet barracks are Thayer, Bartlett, and Mahan halls, which have academic departments, classrooms, laboratories, a computer center, a television studio, three auditoriums, and the Cadet Library. West Point also provides cadets with outstanding athletic facilities and the opportunity to attend religious services in three separate chapels.

West Point Admission Procedure

Applicants must meet certain basic requirements: be a citizen of the United States, at least seventeen and not yet twenty-two years of age on July 1 of the year of admission, and must not be married, pregnant, or have a legal obligation to support a child or children.

Academic qualification: Scholastic transcript, extracurricular record, scores on American College Testing Assessment Program Test (ACT) or College Board Admissions Testing Program Scholastic Aptitute Test (SAT), and recommendations of high school faculty are used to determine this qualification. Leadership, community and extracurricular involvement are also considered.

Physical aptitude qualification: The applicant's Physical Aptitude Examination results, extracurricular record, and recommendations are used to determine physical aptitude qualifications.

Medical qualifications: Applicants must complete a Department of Defense Qualifying Medical Examination, which will be used to determine their medical qualification.

Applicants must also receive a nomination from an authorized source to be considered for appointment to the Military Academy.

Write to Admissions USMA, West Point, NY 10996, so that you can start a West Point file after you complete your precandidate package. Also apply for a nomination from your Congressman. If you are the daughter (or son) of a military person, see your campus counselor about special opportunities that may be open to you.

Women in ROTC

In 1972, as an experimental program, 212 women enrollees in the Reserve Officers Training Corps were accepted at ten colleges and universities. In 1975, for the first time in Army ROTC history, women participated in advanced camp field training along with the men. During the summer of 1976 women ROTC cadets began receiving the same training as men except for necessary variations due to physiological differences. By the school year of 1977–78, 14,296 women were in the program, representing about 24 percent of the total. In the school year of 1975–76, 150 women were commissioned; in 1976–77 there were 495. The number has continued to grow, along with the number of women admitted to West Point.

U.S. ARMY PHOTO

A lieutenant inspects the reserve parachute of a private preparing for the first female parachute jump, at Fort Bragg, North Carolina.

Army Reserve

The Army Reserve traces its traditions back to the Minutemen of Revolutionary days. Today's Reservists are men and women from across the United States who devote some of their spare time to being Reservists.

If you are between seventeen and thirty-four years of age, you may qualify to be a Reservist and gain a well-paying part-time career. With promotions and raises your income can increase, and you can learn valuable skills ranging from aviation to office work. You'll work on community service projects in your hometown and you'll be ready like the Minutemen of two hundred years ago when your country needs you in an emergency.

The Reserve offers opportunities to travel, to meet people, and to learn new skills or perfect old ones. You work sixteen hours a month (usually on weekends) and two weeks each year with your local unit. Your first eight weeks of basic training teach you soldiering. At an

U.S. ARMY PHOTO

A paratrooper after a 1,200-foot jump during the first female parachute jump, at Fort Bragg, North Carolina.

Army post, you build your body and sharpen your mind. You get used to push-ups and parades, but it's worth it as you plan the pattern of your life to help yourself and your country. You also make lifelong friends.

If you're seventeen years or older and join the Reserve as a high school senior, you can start earning over $75 a month nine months before you go for basic training in the summer. You earn over $620 plus room and board each month you're away, so it's better than a summer job pumping gas or waiting on tables.

If you join the Reserve as a junior, you start earning an income three months before basic training. All this gives you a jump on the job market when you graduate from high school.

You can earn over $6,300 during four years of college. You receive $100 a month while you're in college, and you serve only one weekend a month so it doesn't interfere with your schoolwork. Under the Split Training Option you can split your training into two consecutive summers, so you can total over $2,850 for summer training over the four years. The rest of the year you can earn over $75 a month for Reserve duty with your local unit—worth over $3,450 during your four years of college. All that adds up to more than $6,300 in just four years.

For even more, you can join a Simultaneous Membership Program—both Army ROTC and Reserve—while in college. You receive $100 each month for two years in the ROTC advanced course, plus your Reserve pay. After two satisfactory years in the program, you're commissioned a second lieutenant while you're still in college. (This program is not available to ROTC scholarship students.)

Many Reserve units offer a choice of a $2,000 enlistment bonus or up to $5,040 in educational assistance in addition to the extra income you earn.

If you're a high school diploma graduate and choose the enlistment bonus, you receive half when you complete your initial training and the balance over the next six years.

If you decide on the educational assistance, you receive as much as $1,000 for each year of college or vocational or technical school, up to a maximum of $5,040. If you have a National Direct Student Loan or a Guaranteed Student Loan made after October 1, 1975, you can get help in repaying your debt by joining the Army Reserve. For every year you serve in the Reserve, 15 percent of your debt or $500, whichever is greater, is repaid (not to exceed a total of $10,000).

The Army ROTC Two-Year Program

The program is designed for junior and community college graduates, for students at four-year institutions who did not take ROTC in the first two years of college, for students entering a two-year postgraduate course of study, and for students at military junior colleges.

Under this program, you can graduate from college with a diploma and an officer's commission, and you will receive valuable leadership and management training.

The first step is basic camp, a fully paid six-week training camp normally held during the summer between your sophomore and junior years. Rappeling, guiding a group to safety in a test of leadership, swimming with full gear, and navigating in the wilderness using only a map and compass are some of the practical elements the Army includes in basic camp programs. Reflexes are tested, wits are sharpened, and teamwork is learned. The self-confidence you gain will prove even more valuable than the pay, which amounts to more than $600 plus room, board, and transportation. Exceptional performance at basic camp can qualify you for a two-year scholarship that pays most tuition, and educational fees, and a specified amount for books, supplies, and equipment. It also entitles you to a subsistence allowance of up to $1,000 for each of the two years. Even if you don't win a scholarship, you still receive the subsistence allowance for the two years along with your uniform and military science textbooks and material.

With basic camp completed, you become eligible to enter the advanced course. During the next two years in college, you receive advanced instruction in leadership development, organization and management tactics, and ethics and professionalism. Here you develop a sense of responsibility and discipline as you learn business and administrative skills, how to lead people, and how to manage money and equipment—qualifications equally valued in an Army officer and a civilian employee.

As a graduate of the Army ROTC Two-year Program, you may be selected to serve three years on active duty and five in the Reserve Forces, or to serve eight years in Reserve Forces Duty. If the latter, you serve three to six months on active duty as an officer in the basic course and the remainder of your eight-year obligation in the Reserve Forces.

All scholarship graduates may be required to serve three to four years on active duty in the U.S. Army, followed by four to six years in the Reserve Forces.

To enroll, get in touch with the Professor of Military Science at one of the colleges or universities listed at the end of this chapter. He will arrange an orientation for you and a medical examination. You should see him early in the school year to determine your eligibility for basic camp during the following summer.

Army ROTC Scholarships

You can cut the high cost of college with an Army ROTC scholarship. Four-year, three-year, and two-year scholarships are awarded on a competitive basis. The scholarships pay for most college tuition and educational fees, and a specified amount for textbooks, supplies, and equipment. In addition, you receive a subsistence allowance of up to $1,000 each school year the scholarship is in effect. You also receive pay for attending the six-week advanced camp during the summer between your junior and senior years of college.

A large number of these scholarships go to students seeking degrees in engineering, physical sciences, or nursing.

Qualifications and Requirements

To qualify for a four-year Army ROTC scholarship, you must:

Be a citizen of the United States when you accept the award.

Be at least 17 years old before the scholarship becomes effective.

Take either the Scholastic Aptitude Test (SAT) or the American College Test (ACT) *no later than December of the year you apply for the scholarship.*

Have good school grades.

Participate in leadership, extracurricular, and athletic activities. (If you hold a part-time job and do not have time for such activities, substitute credit is given.)

Meet required physical standards.

Be under twenty-five years old on June 30 of the year you expect to graduate and receive your officer's commission. (An age extension of up to four years is possible for veterans who qualify.)

Be a high school graduate or have equivalent credit.

Be accepted by one of the colleges or universities that host Army ROTC.

Pursue a Department of the Army academic discipline.

Successfully complete at least one quarter semester of college instruction in a major Indo-European or Asian language.
Agree to accept a commission as a Regular Army, Army National Guard, or U.S. Army Reserve officer, whichever is offered.

The period for requesting application forms for four-year Army ROTC scholarships is from April 1 through November 15. If your application is received by August 15, you will be considered for early scholarship. Early winners are announced about November 1.

If your application was received after August 15, or if you were not selected by the first board, you will be considered for a later cycle.

For application forms and information, write to:

Army ROTC
PO Box 9000
Clifton, NJ 07015-9974

(You may apply for a QEP four-year scholarship if you wish to attend a historically black college that hosts the Army ROTC program. A limited number of these scholarships are awarded each year on a competitive basis.)

Military Service Obligations

All scholarship graduates must serve in the military for eight years. They may serve three to four years on active duty followed by service in the Army National Guard or U.S. Army Reserve, or serve eight years in the Army National Guard or Reserve preceded by three to six months of active duty. Winners of Army ROTC scholarships are required to maintain acceptable standards of academic achievement, personal conduct, and physical fitness.

If you fail to complete the ROTC program for a commission, you are required by law to serve in enlisted status to compensate the Army for the cost of your scholarship. Your Professor of Military Science is authorized to revoke your scholarship (either temporarily or permanently) if he perceives an insincere commitment or failure to meet the terms involved.

An applicant for active duty in the Army Nurse Corps must be a BSN candidate in the senior semester, a new graduate of a BSN program, or a working nurse with a BSN or higher degree; must be between the ages of twenty-one and thirty-three; must be a citizen of the U.S. or a permanent resident, and must meet height and weight requirements.

An applicant for reserve duty must be a graduate of a diploma program or hold an associate, BSN, or higher degree and have passed the State Board examination; must be between the ages of twenty-one and forty-seven; must be a citizen or permanent resident, and must meet height and weight requirements.

Army Nurse Corps applicants attend an Officers Orientation Course at Fort Sam Houston, Texas. The course is nine weeks for active duty and two weeks for reserve duty. Nurses receive a direct commission as Second Lieutenant.

The Army Nurse Corps offers competitive starting salaries with annual cost-of-living raises, as well as longevity raises every year for the first four years and every two years thereafter. Nurses have an opportunity to continue their education, with 75 percent tuition assistance. They incur a three-year service obligation.

Additional benefits include the following:

- Free medical care for self and dependents.
- Vacation of thirty days per year with full pay.
- Use of military aircraft to almost anywhere in the world.
- "Military discounts" on civilian airlines.
- No loss of seniority when changing locations.
- Retirement after twenty years of service.
- Free dental care.
- Low-cost life insurance.
- Reduced rates at the post exchange (department store) and commissary (grocery store).
- Opportunity to practice nursing in exciting locations all over the United States and overseas.

Army Language Programs

Army language programs are designed by the Defense Language Institute (DLI), including both tapes and books, for either self-study or classroom. They are broken down into four skill levels:

Headstart, a vocabulary of 500 to 800 elementary survival-skill words and acculturation—40 hours of study.

Gateway, designed for officers and noncommissioned officers—500- to 800-word vocabulary emphasizing social, official, and military situations. (DLI Resident)

Basic Courses, from 24 to 47 weeks requiring 144 to 280 hours of study, designed to give a proficiency rating. (DLI Resident)

Refresher or Intermediate, advanced proficiency.

(Among the requirements for these courses, officers must have served three years and be high school graduates. However, the booklets and cassettes can be purchased for a nominal fee by military or civilian individuals from the Defense Language Institute.)

The list of courses available fires the imagination like a trip around the world. They are as follows: *Headstart*, German, Korean, Turkish, Arabic, Spanish for Panama, Japanese; *Video*, Spanish for Puerto Rico; *Gateway*, Korean, German; *Basic*, Arabic, Chinese, Cantonese, Chinese Mandarin, Czech, French, Greek, Hungarian, Italian, Korean, Polish, Portuguese, Romanian, Russian, Serbo-Croatian, Spanish Castilian, Thai, Turkish, Vietnamese; *Special or Short*, German, Portuguese, and advanced Korean, Chinese Mandarin, French, Hebrew, and Russian.

English as a Second Language is also available to soldiers whose primary language is other than English, and on a space-available basis to family members; it can be a valuable tool in making up educational deficiencies.

Army Junior ROTC

The Junior ROTC gives you a chance in your high school years to gain practical experience in organization, map reading, leadership, and teamwork. Military history, weapons safety, and marksmanship and drill are some of the skills you learn. Your uniforms, textbooks, and equipment are provided by the Army and your school, at no expense to you or your family. Your enrollment in the program does not obligate you for any future military service. While you develop good citizenship and self-reliance physically and mentally, you may enjoy JROTC extracurricular activities such as military bands and drum and bugle corps, social events such as picnics and barbecues, and at some schools a military ball with the crowning of a queen.

As to your future, JROTC can help you if you choose to compete later for an Army ROTC four-year scholarship.

How You Can Earn Up to $25,200 for College

With the Montgomery GI Bill plus the Army College Fund, you can earn up to $25,200 while you serve your country.

Here's what the Army says:

"First, you contribute to your education—$100 per month for the first twelve months of your enlistment. Then the government contributes—up to $9,600. That's the new GI Bill. Your contributions will be deducted automatically from your Army pay.

"The Montgomery GI Bill is available to everyone who enlists in the Army after July 1, 1985, regardless of what specialty you sign up for. You cannot have served during a previous Army enlistment. How much you'll earn depends on how long you serve. For maximum returns, see the table below.

"But you can accumulate up to an additional $14,400 through the Army College Fund. This is the Army's way to encourage ambitious high school graduates to think seriously about Army service.

"To qualify for the extra benefits, you must meet certain standards. You must be a high school graduate, score 50 or more on the Armed Forces Qualification Test, and enlist for a selected Army specialty.

Enlistment	*New GI Bill*	*New GI Bill Plus New ACF*
2 years	$ 9,000	$17,000
3 years	$10,800	$22,800
4 years	$10,800	$25,200

ARMY ROTC—Colleges and Universities

ALABAMA

Alabama A&MUniversity, Normal
Auburn Univ., Auburn
***Auburn Univ. at Montgomery, Montgomery
Jacksonville State Univ., Jacksonville

**Marion Military Institute, Marion
Tuskegee Institute, Tuskegee
Univ. of Alabama, University
Univ. of Alabama, Birmingham, Birmingham
Univ. of North Alabama, Florence
Univ. of South Alabama, Mobile

ALASKA

Univ. of Alaska-Fairbanks, Fairbanks

ARIZONA

Arizona State Univ., Tempe
Northern Arizona Univ., Flagstaff
Univ. of Arizona, Tucson

ARKANSAS

Arkansas State Univ., State University
Arkansas Tech Univ., Russellville
Henderson State Univ., Arkadelphia
Ouachita Baptist Univ., Arkadelphia
Southern Arkansas Univ., Magnolia
Univ. of Arkansas, Fayetteville
Univ. of Arkansas, Little Rock, Little Rock
Univ. of Arkansas at Pine Bluff, Pine Bluff
Univ. of Central Arkansas, Conway

CALIFORNIA

California Polytechnic State Univ., San Luis Obispo
***California State College-San Bernardino, San Bernardino
***California State Polytechnic Univ.-Pomona, Pomona
California State Univ. at Fresno, Fresno
California State Univ. at Long Beach, Long Beach
***California State Univ.-Fullerton, Fullerton
San Diego State Univ., San Diego
San Jose State Univ., San Jose
The Claremont Colleges, Claremont
Univ. of California-Berkeley, Berkeley
***California State Univ. at Chico
Univ. of California-Davis, Davis
Univ. of California-Los Angeles, Los Angeles
Univ. of California-Santa Barbara, Santa Barbara
***California State Univ. at Sacramento
Univ. of San Francisco, San Francisco
Univ. of Santa Clara, Santa Clara
Univ. of Southern California, Los Angeles

COLORADO

Colorado School of Mines, Golden
Colorado State Univ., Fort Collins
***Mesa College, Grand Junction
Metropolitan State College, Denver
Univ. of Colorado, Boulder
Univ. of Colorado at Colorado Springs, Colorado Springs
Univ. of Southern Colorado, Pueblo

CONNECTICUT
***Greater Hartford Campus, Univ. of Connecticut, Hartford
Univ. of Connecticut, Storrs
***Univ. of Bridgeport, Bridgeport

DELAWARE
Univ. of Delaware, Newark

DISTRICT OF COLUMBIA
Georgetown Univ., Washington
Howard Univ., Washington

FLORIDA
Embry-Riddle Aeronautical Univ., Daytona Beach
Florida A&M Univ., Tallahassee
Florida Institute of Technology, Melbourne
Florida Southern College, Lakeland
Florida State Univ., Tallahassee
***Saint Leo College, Saint Leo
Stetson Univ., DeLand
***Univ. of Central Florida, Orlando
Univ. of Florida, Gainesville
Univ. of Miami, Coral Gables
***Univ. of North Florida, Jacksonville
***Univ. of South Florida-St. Petersburg, St. Petersburg
Univ. of South Florida, Tampa
Univ. of Tampa, Tampa
***Univ. of West Florida, Pensacola

GEORGIA
***Albany State College, Albany
***Armstrong State College, Savannah
Augusta State College, Augusta
***Berry College, Mount Berry
Columbus College, Columbus
Fort Valley State College, Fort Valley
Georgia Institute of Technology, Atlanta
**Georgia Military College, Milledgeville
Georgia Southern College, Statesboro
***Georgia Southwestern College, Americus
Georgia State Univ., Atlanta
Mercer Univ., Macon
*North Georgia College, Dahlonega
Univ. of Georgia, Athens

GUAM
Univ. of Guam, Agana

HAWAII
Univ. of Hawaii, Honolulu

IDAHO
Boise State Univ., Boise
Idaho State Univ., Pocatello
Univ. of Idaho, Moscow

ILLINOIS

***Bradley Univ., Peoria
Chicago State Univ., Chicago
Eastern Illinois Univ., Charleston
Illinois State Univ., Normal
Knox College, Galesburg
Loyola Univ. of Chicago, Chicago
Northern Illinois Univ., DeKalb
Southern Illinois Univ., Carbondale
Univ. of Illinois, Urbana-Champaign
Univ. of Illinois-Chicago Circle, Chicago
Western Illinois Univ., Macomb
Wheaton College, Wheaton

INDIANA

Ball State Univ., Muncie
Indiana Univ., Bloomington
Indiana Univ.-Purdue Univ. at Indianapolis, Indianapolis
***Indiana Univ.-Southeast, New Albany
Purdue Univ., West Lafayette
Rose-Hulman Institute of Technology, Terre Haute
Univ. of Notre Dame, Notre Dame

IOWA

***Drake Univ., Des Moines
Iowa State Univ. of S&T, Ames
***Univ. of Dubuque, Dubuque
Univ. of Iowa, Iowa City
***Univ. of Northern Iowa, Cedar Falls

KANSAS

***Emporia State Univ., Emporia
***Fort Hays State Univ., Fort Hays
Kansas State Univ. of A&AS, Manhattan
Pittsburg State Univ., Pittsburg
Univ. of Kansas, Lawrence
Wichita State Univ., Wichita

KENTUCKY

***Cumberland College, Williamsburg
Eastern Kentucky Univ., Richmond
***Kentucky State Univ., Frankfort
Morehead State Univ., Morehead
Murray State Univ., Murray
***Northern Kentucky Univ., Highland Heights
Univ. of Kentucky, Lexington
Univ. of Louisville, Louisville
Western Kentucky Univ., Bowling Green

LOUISIANA

***Centenary College of Louisiana, Shreveport
***Dillard Univ., New Orleans

***Grambling State Univ., Grambling
Louisiana State Univ. and A&M College, Baton Rouge
***Louisiana State Univ. at Shreveport, Shreveport
McNeese State Univ., Lake Charles
Nicholls State Univ., Thibodaux
Northeast Louisiana Univ., Monroe
Northwestern State Univ. of Louisiana, Natchitoches
Southeastern Louisiana Univ., Hammond
Southern Univ. and A&M College, Baton Rouge
Tulane Univ., New Orleans

MAINE

Univ. of Maine, Orono
Univ. of Southern Maine, Portland

MARYLAND

***Bowie State College, Bowie
***Frostburg College, Frostburg
Loyola College, Baltimore
Morgan State Univ., Baltimore
***Mount Saint Mary's College, Emmitsburg
***Salisbury State College, Salisbury
The Johns Hopkins Univ., Baltimore
Western Maryland College, Westminster

MASSACHUSETTS

Boston Univ., Boston
***Fitchburg State College, Fitchburg
Massachusetts Institute of Technology, Cambridge
Northeastern Univ., Boston
***Salem State College, Salem
***Stonehill College, North Easton
***Suffolk Univ., Boston
Univ. of Massachusetts, Amherst
***Western New England College, Springfield
Worcester Polytechnic Institute, Worcester

MICHIGAN

Central Michigan Univ., Mount Pleasant
Eastern Michigan Univ., Ypsilanti
Michigan State Univ., East Lansing
Michigan Technological Univ., Houghton
Northern Michigan Univ., Marquette
Univ. of Detroit, Detroit
Univ. of Michigan, Ann Arbor
Western Michigan Univ., Kalamazoo

MINNESOTA

Bemidji State Univ., Bemidji
Mankato State Univ., Mankato
St. John's Univ., Collegeville
Univ. of Minnesota, Minneapolis
Winona State Univ., Winona

MISSISSIPPI

Alcorn State Univ., Lorman
***Copiah-Lincoln Jr. College, Wesson
Delta State Univ., Cleveland
Jackson State Univ., Jackson
***Jefferson Davis Jr. College, Gulfport
***Meridian Junior College, Meridian
Mississippi State Univ., Mississippi State
***Southwest Mississippi Jr. College, Summit
Univ. of Mississippi, University
Univ. of Southern Mississippi, Hattiesburg

MISSOURI

Central Missouri State Univ., Warrensburg
***Evangel College, Springfield
**Kemper Military School and College, Boonville
Lincoln Univ., Jefferson City
***Missouri Southern State College, Joplin
Missouri Western State College, St. Joseph
Northeast Missouri State Univ., Kirksville
Northwest Missouri State Univ., Maryville
***Southeast Missouri State Univ., Cape Girardeau
Southwest Missouri State Univ., Springfield
Univ. of Missouri-Columbia, Columbia
Univ. of Missouri-Rolla, Rolla
***Univ. of Missouri-St. Louis, St. Louis
Washington Univ., St. Louis
**Wentworth Military Academy and Junior College, Lexington
Westminster College, Fulton

MONTANA

***Eastern Montana College, Billings
Montana State Univ., Bozeman
Univ. of Montana, Missoula

NEBRASKA

Creighton Univ., Omaha
Kearney State College, Kearney
Univ. of Nebraska, Lincoln
***Univ. of Nebraska-Omaha, Omaha

NEVADA

Univ. of Nevada-Las Vegas, Las Vegas
Univ. of Nevada, Reno

NEW HAMPSHIRE

Univ. of New Hampshire, Durham
***Dartmouth College, Hanover

NEW JERSEY

***Jersey City State College, Jersey City
***Monmouth College, West Long Beach

Princeton Univ., Princeton
Rider College, Lawrenceville
Rutgers Univ., New Brunswick
Seton Hall Univ., South Orange
St. Peter's College, Jersey City

NEW MEXICO

Eastern New Mexico Univ., Portales
**New Mexico Military Institute, Roswell
New Mexico State Univ., Las Cruces
***Univ. of Albuquerque, Albuquerque

NEW YORK

Canisius College, Buffalo
Clarkson College of Technology, Potsdam
Cornell Univ., Ithaca
Fordham Univ., Bronx
Hofstra Univ., Hempstead
***John Jay College of Criminal Justice, City Univ. of New York, New York City
***Marist College, Bronx
Niagara Univ., Niagara University
Polytechnic Institute of New York, Brooklyn
Rensselaer Polytechnic Institute, Troy
Rochester Institute of Technology, Rochester
Siena College, Loudonville
St. Bonaventure Univ., St. Bonaventure
St. Johns Univ., Jamaica
St. Lawrence Univ., Canton
***State Univ. of New York at Albany, Albany
State Univ. of New York at Brockport, Brockport
***State Univ. of New York at Cortland, Cortland
State Univ. of New York at Fredonia, Fredonia
***State Univ. of New York at Oswego, Oswego
Syracuse Univ., Syracuse

NORTH CAROLINA

Appalachian State Univ., Boone
Campbell Univ., Buies Creek
Davidson College, Davidson
Duke Univ., Durham
***East Carolina Univ., Greenville
***Elizabeth City State Univ., Elizabeth City
***Elon College, Elon College
North Carolina A&T State Univ., Greensboro
North Carolina State Univ. at Raleigh, Raleigh
St. Augustine's College, Raleigh
***Univ. of North Carolina at Charlotte, Charlotte
Univ. of North Carolina at Wilmington, Wilmington
Wake Forest Univ., Winston-Salem
Western Carolina Univ., Cullowhee

NORTH DAKOTA

North Dakota State Univ. of A&AS, Fargo
Univ. of North Dakota, Grand Forks

OHIO

Bowling Green State Univ., Bowling Green
Central State Univ., Wilberforce
***Franklin Univ., Columbus
John Carroll Univ., Cleveland
Kent State Univ., Kent
Ohio State Univ., Columbus
Ohio Univ., Athens
***Rio Grande College, Rio Grande
Univ. of Akron, Akron
Univ. of Cincinnati, Cincinnati
Univ. of Dayton, Dayton
Univ. of Toledo, Toledo
***Wright State Univ., Dayton
Xavier Univ., Cincinnati
Youngstown State Univ., Youngstown

OKLAHOMA

Cameron Univ., Lawton
Central State Univ., Edmond
East Central Oklahoma State Univ., Ada
Northwestern Oklahoma State Univ., Alva
Oklahoma State Univ., Stillwater
Southwestern Oklahoma State Univ., Weatherford
Univ. of Oklahoma, Norman
***Univ. of Tulsa, Tulsa

OREGON

***Eastern Oregon State College, LaGrande
***Oregon Institute of Technology, Klamath Falls
Oregon State Univ., Corvallis
***Portland State Univ., Portland
Univ. of Oregon, Eugene

PENNSYLVANIA

***Altoona Campus, Penn State Univ., Altoona
***Behrend College, Penn State Univ., Erie
Bucknell Univ., Lewisburg
***California Univ. of Pennsylvania, California
Carnegie-Mellon Univ., Pittsburgh
Clarion Univ. of Pennsylvania, Clarion
***Delaware County Campus, Penn State Univ., Media
Dickinson College, Carlisle
Drexel Univ., Philadelphia
Duquesne Univ., Pittsburgh
***East Stroudsburg Univ. of Pennsylvania, East Stroudsburg
***Edinboro State Univ. of Pennsylvania, Edinboro
Gannon Univ., Erie
Gettysburg College, Gettysburg
***Hazelton Campus, Penn State Univ., Hazelton
Indiana Univ. of Pennsylvania, Indiana
Lafayette College, Easton

LaSalle College, Philadelphia
Lehigh Univ., Bethlehem
***Lock Haven Univ. of Pennsylvania, Lock Haven
***Mansfield Univ. of Pennsylvania, Mansfield
***Millersville Univ. of Pennsylvania, Millersville
***Ogontz Campus, Penn State Univ., Abington
Pennsylvania State Univ., University Park
***Schuylkill Campus, Penn State Univ., Schuylkill Haven
Shippensburg Univ. of Pennsylvania, Shippensburg
***Slippery Rock Univ. of Pennsylvania, Slippery Rock
Temple Univ., Philadelphia
Univ. of Pennsylvania, Philadelphia
Univ. of Pittsburgh, Pittsburgh
Univ. of Scranton, Scranton
**Valley Forge Military Academy and Junior College, Wayne
Washington and Jefferson College, Washington
Widener College, Chester

PUERTO RICO

Univ. of Puerto Rico, Mayaguez Campus, Mayaguez
Univ. of Puerto Rico, Rio Piedras Campus, Rio Piedras

RHODE ISLAND

***Bryant College, Smithfield
Providence College, Providence
***Rhode Island College, Providence
Univ. of Rhode Island, Kingston

SOUTH CAROLINA

***Benedict College, Columbia
Clemson Univ., Clemson
***Francis Marion College, Florence
Furman Univ., Greenville
Presbyterian College, Clinton
South Carolina State College, Orangeburg
*The Citadel, Charleston
Univ. of South Carolina, Columbia
Wofford College, Spartanburg

SOUTH DAKOTA

***Black Hills State College, Spearfish
***Northern State College, Aberdeen
South Dakota School of Mines and Technology, Rapid City
South Dakota State Univ., Brookings
Univ. of South Dakota, Vermillion

TENNESSEE

Austin-Peay State Univ., Clarksville
Carson-Newman College, Jefferson City
East Tennessee State Univ., Johnson City
Memphis State Univ., Memphis
Middle Tennessee State Univ., Murfreesboro

Tennessee Technological Univ., Cookeville
Univ. of Tennessee, Knoxville
Univ. of Tennessee at Chattanooga, Chattanooga
Univ. of Tennessee at Martin, Martin
Vanderbilt Univ., Nashville

TEXAS
Hardin-Simmons Univ., Abilene
***Howard Payne Univ., Brownwood
***Lamar Univ., Beaumont
Midwestern State Univ., Wichita Falls
Pan American Univ., Edinburg
Prairie View A&M Univ., Prairie View
Sam Houston State Univ., Huntsville
Stephen F. Austin State Univ., Nacogdoches
St. Mary's Univ., San Antonio
***Tarleton State Univ., Stephenville
Texas A&I Univ., Kingsville
Texas A&M Univ., College Station
Texas Christian Univ., Fort Worth
Texas Tech Univ., Lubbock
***Texas Woman's Univ., Denton
Trinity Univ., San Antonio
Univ. of Houston, Houston
Univ. of Texas at Arlington, Arlington
Univ. of Texas at Austin, Austin
Univ. of Texas at El Paso, El Paso
Univ. of Texas at San Antonio, San Antonio
West Texas State Univ., Canyon

UTAH
Brigham Young Univ., Provo
Univ. of Utah, Salt Lake City
Utah State Univ., Logan
Weber State College, Ogden

VERMONT
*Norwich Univ., Northfield
Univ. of Vermont, Burlington

VIRGINIA
***Christopher Newport College, Newport News
***George Mason Univ., Fairfax
Hampton Institute, Hampton
James Madison Univ., Harrisonburg
***Longwood College, Farmville
***Lynchburg College, Lynchburg
Norfolk State Univ., Norfolk
Old Dominion Univ., Norfolk
The College of William and Mary, Williamsburg

Univ. of Richmond, Richmond
Univ. of Virginia, Charlottesville
*Virginia Military Institute, Lexington
Virginia Polytechnic Institute and State Univ., Blacksburg
Virginia State Univ., Petersburg
Washington and Lee Univ., Lexington

WASHINGTON

***Central Washington Univ., Ellensburg
Eastern Washington Univ., Cheney
Gonzaga Univ., Spokane
Seattle Univ., Seattle
Univ. of Washington, Seattle
Washington State Univ., Pullman

WEST VIRGINIA

Marshall Univ., Huntington
West Virginia State College, Institute
West Virginia Univ., Morgantown

WISCONSIN

Marquette Univ., Milwaukee
Ripon College, Ripon
St. Norbert College, DePere
Univ. of Wisconsin-LaCrosse, LaCrosse
Univ. of Wisconsin-Madison, Madison
Univ. of Wisconsin-Milwaukee, Milwaukee
Univ. of Wisconsin-Oshkosh, Oshkosh
Univ. of Wisconsin-Platteville, Platteville
Univ. of Wisconsin-Stevens Point, Stevens Point
Univ. of Wisconsin-Whitewater, Whitewater

WYOMING

Univ. of Wyoming, Laramie

---

*Military College or University
**Military Junior College
***Extension Center

Army Occupations

Career fields are listed alphabetically, beginning with those combat-related groups closed to women (indicated by "C"). Those marked with an asterisk (*) are generally open to all applicants, but may include specific combat-related positions open only to men.

Career Fields	Duties & Responsibilities	Qualifications	Examples of Civilian Jobs
Armor (C)	Performs combat tasks using tanks and armored reconnaissance vehicles.	Team sports, mechanical maintenance, orienteering.	Heavy equipment operator, supervisor.
Combat (C) Engineering	Constructs and maintains roads and bridges; operates powered bridges; constructs and clears minefields and other personnel and vehicular obstacles, demolitions with high explosives; erects temporary shelters and sets up camouflage.	Automotive mechanics, carpentry, woodworking, mechanical drawing and drafting courses.	Blaster, construction equipment operator, construction supervisor, bridge repairer and lumber worker.
Infantry (C)	Performs combat tasks using rifles, mortars, tank destroying missiles, personnel carriers, vehicle-mounted guns and fire control equipment.	Team sports, orienteering, hunting and other outdoor sports.	Supervisor, gunsmith, security officer, firearms handler.
Administration	Performs general administrative duties such as typing, stenography and postal functions and specialized administrative duties such as personnel, legal, equal opportunity and chapel duties.	Basic clerical and communication abilities, typing, bookkeeping, stenography or office management skills desirable.	Clerk typist, secretary, employment interviewer, postal clerk, recreation specialist, office manager, personnel clerk, bookkeeper, cashier, payroll clerk, court clerk.

Career Fields	Duties & Responsibilities	Qualifications	Examples of Civilian Jobs
Aircraft Maintenance	Performs the mechanical functions of maintenance, repair and modification of helicopters, turbo-prop, and reciprocating engine aircraft.	Considerable mechanical or electrical aptitude and manual dexterity. Shop mathematics and physics desirable.	Aircraft mechanic, plane inspector.
Aircraft System Maintenance	Performs maintenance of aircraft accessory systems, propulsion systems, armament systems, fabrication of metal materials used in aircraft structural repair and inspection and preservation.	Electrical and mechanical aptitude, shop mathematics and shop work is desirable.	Aircraft mechanic, aircraft electrician, sheet metal worker machinist.
Aircrew Protection	Receives, stores, issues and maintains aircrew survival and protective equipment.	High electrical and mechanical aptitude, manual dexterity, normal vision. Mathematics, physics and shop work desirable.	No civilian job covers the scope of duties in this career field.
Air Defense Artillery (*)	Emplaces, assembles, tests, maintains and fires air defense weapons systems; operates fire control equipment, radars, computers, automatic data transmission and associated power supply equipment.	Basic mechanical, electrical, electronic mathematical abilities; emotional stability and high degree of reasoning ability.	Map and topographical drafter, cartographer.
Air Defense Missile Maintenance	Inspects, tests, maintains and repairs guided missile fire control equipment and related radar installations which guide missiles to target.	Mathematics, physics, electricity and electronics.	Radio installation and repair inspector, electronic equipment technician, radio and TV repairer.

Ammunition	Handles, stores, reconditions and salvages ammunition, explosives, and components; locates, removes and destroys or salvages unexploded bombs and missiles.	Mechanical aptitude, attentiveness, good close vision, normal color discrimination, manual dexterity and hand-eye coordination.	Toxic chemical handler, ammunition inspector and acid plant operator.
Automatic Data Processing	Operates a variety of electric accounting and automatic data processing equipment to produce personnel, supply, fiscal, medical, intelligence and other reports.	Reasoning and verbal ability, clerical aptitude, finger and manual dexterity and hand-eye coordination. Knowledge of typing and office machines.	Coding clerk, key punch, computer and sorting machine operator, machine records unit supervisor.
Aviation Communications—Electronic Systems Maintenance	Repairs and maintains navigation, flight control, ground control approach radar, aerial surveillance and associated communications equipment.	Electrical/electronic theory and repair.	Electronics technician, radar repairer, electrical instrument machine repairer.
Aviation Maintenance	Performs general maintenance on fixed-wing and rotary-wing aircraft. Operates and maintains aircraft weapon systems and serves as flying crew chief.	Aircraft and automotive mechanics, blueprint reading, electricity, sheet metal working, mathematics and physics.	Aircraft engine mechanic, airframe repairer, airplane electrician and aircraft fuel systems mechanic.
Avionics	Installs and/or repairs aircraft radio and radar systems.	Mathematics and shop courses in electricity and electronics useful.	Radio and television or electrical instrument repairer, communications electrical and electronics engineer and radio engineer.
Ballistic/Land Combat Missile and Light Air Defense Weapons System Maintenance	Inspects, tests, maintains and repairs tactical missile systems and related test equipment and trainers.	Mathematics, physics, electricity, electronics (radio and TV) and blueprint reading.	Electronic equipment technician, radio electrician and mechanic, TV repair and service technician.

Career Fields	Duties & Responsibilities	Qualifications	Examples of Civilian Jobs
Band	Plays brass, woodwind or percussion instrument in marching, concert, dance, stage and show bands, combos or instrumental ensembles. Sings in vocal group, writes and arranges music.	Instrumental audition on brass, woodwind or percussion instrument.	Bandperson, bandmaster, musician, accompanist, arranger, music director, orchestrator, music teacher, orchestra leader and vocalist.
Chemical	Provides decontamination service after chemical, biological or radiological attacks, produces smoke for battlefield concealment, repairs chemical equipment and assists in overall planning of chemical, biological, or radiological activities.	Biology, chemistry and electricity.	Laboratory assistant (biological, chemical or radiological), pumper and repairer (chemical) and exterminator.
Communications—Electronics Maintenance	Installs, maintains radar and radio receiving, transmitting carrier and terminal equipment.	Electricity, mathematics, electronics and blueprint reading.	Radio control room technician, radio mechanic, transmitter, radio and TV repairer.
Communications—Electronics Operations	Installs, maintains field telephone switchboards and field radio communications equipment.	Mathematics, physics and shop courses in electricity.	Communications engineer assistant, plant electrician and radio electrician or operator.
Electronic Warfare Cryptologic Operations	Collects and analyzes electromagnetic emissions; ensures communications security; performs electronic warfare duties in fixed or mobile operations.	Verbal and reasoning ability and perceptual speed; aural and visual acuity.	Radio and telegraph operator, navigator, intelligence research analyst, statistician, signal collection technician.
Electronic Warfare Intercept Systems Maintenance	Installs, operates and maintains intercept, electronic measuring and testing equipment.	Physics, mathematics, electricity, electronics (radio and TV) and blueprint reading.	Electrical instrument, meteorological instrument repairer and electronic equipment inspector.

Finance and Accounting	Maintains pay records of military personnel; prepares vouchers for payment; prepares reports, disburses funds; accounts for funds, to include budgeting, allocation, auditing; compiles and analyzes statistical data and prepares cost analysis records.	Dexterity in the operation of business machines. Typing, mathematics, statistics and basic principles of accounting desirable. High administrative aptitude mandatory.	Paymaster, cashier, statistical or audit clerk, accountant, budget clerk and bookkeeper.
Field Artillery (*)	Operates, maintains and directs fire or field artillery guns, howitzers, missiles, rockets and related weapons. Operates and maintains supporting equipment such as target acquisition radars, sound and flash ranging, meteorological and survey equipment.	Emotional stability, mathematics and reasoning abilities.	Map and topographical drafter, cartographer, surveyor, weather chart preparer.
Food Service	Plans regular and special diet menus, cooks and bakes food in dining facilities and during field exercises. Serves as aide and cook on personal staff of general officer.	Home economics, work in a restaurant, bake shop or meat market.	Cook, chef, caterer, baker, butcher, kitchen supervisor and cafeteria manager.
General Engineering	Provides utilities and engineering services such as electric power production, building and roadway construction and maintenance, salvage activities, airstrip construction, firefighting and crash rescue operations.	Mechanical aptitude, emotional stability and ability to visualize spatial relationships. Carpentry, woodworking or mechanical drawing.	Carpenter, construction equipment operator, electrician, firefighter, driver, plumber, welder, bricklayer.

Career Fields	Duties & Responsibilities	Qualifications	Examples of Civilian Jobs
Law Enforcement	Enforces military regulations; protects facilities, roads and designated sensitive areas and personnel; controls traffic movement; guards military prisoners and enemy prisoners of war; investigates traffic accidents and crimes involving military personnel.	Sociology and demonstrated prowess and leadership in athletics and other group work helpful.	Police officer, plant guard, detective, investigator, crime detection laboratory assistant and ballistics expert.
Mechanical Maintenance	Services and repairs land and amphibious wheel and track vehicles ranging from cars and light trucks to heavy tanks and self-propelled weapons; installs and repairs refrigeration, bakery and laundry equipment.	Automotive mechanics, electricity, blueprint reading, machine shop and physics.	Automotive mechanic, motor analyst, bakery, or refrigeration equipment, repairer, frame, wheel alignment and tractor mechanic.
Medical	Assists or supports physicians, surgeons, nurses, dentists, psychologists, social workers, and veterinarians in 29 separate job classifications. Some provide direct patient care in hospitals and clinics, others make and repair eyeglasses, dentures, or ortho-medical equipment, or maintain medical records. All play significant roles in a modern worldwide health care delivery system.	Biology, chemistry, hygiene, sociology, general math, algebra, animal care; knowledge of mechanics of electronics; general clerical skills.	Social worker (case aide), practical nurse, nurse's aide, dental assistant, surgeon's assistant, psychological aid, hospital attendant or orderly, veterinary assistant, food quality control, medical equipment repairer, medical or dental laboratory technician, physical therapy assistant, dietetic technician.

Military Intelligence	Gathers, translates, correlates and interprets information, including photographs, associated with military plans and operations.	English composition, typing foreign languages, economics, geography and history.	Investigator, interpreter, cartographic aid, records analyst, research worker and intelligence analyst (government).
Petroleum	Receives, stores, preserves and distributes bulk packaged petroleum products; performs standard physical and chemical tests of petroleum products.	Hygiene, biology, physics, chemistry and mathematics.	Biological laboratory assistant, petroleum tester, chemical laboratory assistant.
Public Affairs and Audiovisual	Prepares newspaper, radio and television communications; maintains radio and television equipment; performs still and motion picture photography, audiovisual equipment repair; produces graphic illustrations.	Clerical aptitude, emotional stability, auditory acuity, color vision, manual dexterity.	Electronics mechanic, television repairer, reporter, editor, script writer, announcer, sign painter, illustrator.
Safety	Conducts both air and ground safety programs, surveys areas and activities to eliminate hazards, analyzes accident causes and trends.	Knowledge of industrial hygiene, safety education, safety psychology, typing, English and public speaking desirable.	Safety inspector and instructor.
Supply and Service	Receives, stores and issues individual, organizational and expendable supplies, equipment and spare parts; establishes, posts and maintains stock records; repairs and alters textile, canvas and leather supplies, rigs parachutes, decontaminates materials. Performs grave registration functions.	Mathematical ability and perceptual speed in scanning and checking supply documents. Verbal ability; courses in bookkeeping, typing and office machine operation.	Inventory clerk, stock control clerk or supervisor, shipping or parts clerk, warehouse manager, parachute rigger and funeral attendant.

Career Fields	Duties & Responsibilities	Qualifications	Examples of Civilian Jobs
Topographic Engineering	Performs land survey; produces construction drawings and plans, maps, charts, diagrams and illustrated material; constructs scale models of terrain and structures. Operates offset duplicators, presses and bindery equipment. Repairs survey instruments and reproduction equipment.	Mechanical drawing and drafting, blueprint reading, commercial art, fine art, geography and mathematics.	Drafting (structural, mechanical and topographical), cartographic and art layout, model maker, commercial artist.
Transportation	Operates and performs preventive maintenance on passenger, light, medium and heavy cargo vehicles; operates and maintains marine lighterage and harbor craft; performs as air traffic controller.	Mechanical aptitude, manual dexterity, hand-eye coordination, FAA certification for air traffic control, license for vehicle operation.	Truck driver, FAA air traffic controller.

Chapter III

The United States Navy and Marine Corps

> "I must go down to the seas again,
> to the lonely sea and the sky,
> And all I ask is a tall ship and
> a star to steer her by..."
>
> "Sea Fever," — John Masefield

It was on October 13, 1775, that the United States Navy was founded, and on November 10 that the Marine Corps was established. If you want to know what America was like in that day, visit Williamsburg, Virginia, which has been restored to its Colonial appearance, with candle and wig makers and silver shops reminiscent of Paul Revere.

The job of the Navy is to protect the United States by war at sea if necessary, and in the time of John Paul Jones it was certainly necessary. The Marine Corps provides Fleet Marine Forces of combined arms, together with supporting aviation components, for service with the U.S. Navy fleet as landing forces in amphibious operations.

By Act of Congress on April 30, 1798, the Department of the Navy and the office of the Secretary of the Navy were created. In 1949 the National Security Act created the Department of Defense. The Secretary of the Navy (appointed by the President, who is Commander in Chief of the Armed Forces) reports to the Secretary of Defense. The Navy Department includes the U.S. Coast Guard when it is operating as a service in the Navy.

Today the responsibility of the Navy and the Marine Corps is to organize, train, and equip forces for combat at sea so that the seas may remain free. The eyes and ears of the Navy—submarines, aircraft, sonar, radar, and modern communication technology—keep track of what's going on around the world to safeguard citizens of the United States. The Navy is also responsible for protecting other shipping. The Marines are skilled in amphibious operations; that is, operations involving both land and sea.

Although the Navy has fleets in several oceans and mobile units that can go to areas where trouble arises, it also needs land-based supporting forces, so the country is divided into naval districts. Since 1978 women have been eligible for sea duty, but if you enlist in the Navy, there's no guarantee you'll go to sea. You may have a shore job.

Navy Organization

The Department of the Navy is composed of three principal segments: the Navy Department, Shore Establishment, and Operating Force. The Navy Department is made up of assistants to the Secretary of the Navy; other ranking officials directing essential programs; the Chief of Naval Operations; and the Commandant of the Marine Corps.

The Chief of Naval Operations is the senior naval adviser to the Secretary of the Navy and is responsible for the operating forces, determining their needs for personnel, ships, aircraft, weapons systems, and materials. He advises the President and Secretary of Defense on naval matters. He is the Navy member of the Joint Chiefs of Staff.

The Shore Establishment comprises systems commands dealing with space and naval warfare, sea, air, supply, medical, education and training, legal service, facilities engineering, intelligence, and telecommunications. Shore activities include submarine and amphibious bases, air stations, shipyards, and laboratories. The Navy conducts research and development in many fields of science, medicine, and technology.

Overseas commands such as U.S. Naval Forces Europe, Southern and Central Commands, Special Warfare, and Military Sealift Commands comprise the Operating Forces. Also included are the Fleet Marine Forces Atlantic and Pacific and their operating forces. The two fleets operating off the U.S. east and west coasts are the Second and Third Fleets, respectively. The Sixth and Seventh Fleets operate in the Mediterranean and Caribbean Seas and the Pacific and Indian Oceans, respectively.

Navy and Marine Education Programs

The Navy and Marine Corps, like the other services, believe in education—not just basic training, but continuing education through correspondence courses and even graduate education. Also

U.S. NAVY PHOTO

A midshipman sits at the controls of a Sea King helicopter during a training program of the Naval Reserve Officers Training Corps.

available are scholarship programs for two-year courses, four-year courses, and Reserve Officer Training at many universities throughout the United States. They are listed later in this chapter. And if you are qualified, you may well be interested in the Naval Academy at Annapolis. The first women graduated from Annapolis in 1980, having entered in 1976.

Navy Officer Procurement Programs

The Naval Reserve Officer Training Corps (NROTC) offers two basic programs leading to a rewarding career as a Navy or Marine Corps officer: the Scholarship Program and the College (non-scholarship) Program. Each provides a two-year or four-year course of study.

If you are reasonably sure you want to make the Navy or Marine Corps your career, you should apply for the Scholarship Program. Selection is made by a national committee under the quota allotted to your state, and you must be accepted by the NROTC.

As a scholarship student you attend campus drills, naval science classes, and summer training sessions while earning your bachelor's degree. You receive, at government expense and for a maximum of four academic years, tuition, textbooks, and special fees. You also receive a monthly tax-free subsistence of $100 up to forty months, plus uniforms. You are required to complete three summer training periods, for which you receive pay of $100 a month.

Upon completion of naval science and bachelor's degree requirements, you transfer from Reserve status to active duty and receive your commission as a regular Navy or Marine Corps officer. You are then obligated to serve for at least four years.

A Navy ROTC Scholarship Program provides a two-year scholarship for selected young men and women who are matriculated college sophomores. You must have an overall grade average of C+ or better, and a major in mathematics, physical sciences, or one of the engineering disciplines is preferred. You must complete a six-week course of instruction at the Naval Science Institute, Newport, Rhode Island, during the summer before your junior year and attend any other summer training and drills prescribed before your graduation and commission. Students in this two-year program receive full tuition, $100 a month subsistence, and textbooks and special fees during their junior and senior years at an NROTC college or university. The service obligation is four years.

The United States Naval Academy

Founded in 1845 as the Naval School and reorganized in 1850–51 as the Naval Academy, on July 6, 1976, Annapolis admitted its first women, eighty-one in all, to train as Naval and Marine officers.

Annapolis is the capital of Maryland and a storied town indeed, with shops and antique houses and views of the waterfront and sailing ships along the way. The Naval Academy occupies 329 acres on the Severn River. John Paul Jones is buried in the crypt of the lovely domed chapel, and if that brave sailor of long ago could come back and wander these tree-shaded grounds, he would be amazed at how much the Navy of his era has grown, for his were the days when sailors were paid with grog and combat at sea was at close hand by boarding the deck of the enemy vessel.

The center for daily living is Bancroft Hall, one of the largest dormitory complexes in the world. It houses the entire Brigade of Midshipmen, approximately 4,300 young men and women, and contains the Midshipmen's Wardroom, the huge dining hall, and the Nimitz Library, overlooking the river. The library has room for 650,000 books, and computers and educational television are part of the air-conditioned academic complex.

The Robert Crown Center is the home of the Intercollegiate Sailing Hall of Fame, as the academy has a sailing program.

Here also is an activity center with an indoor ice-skating rink, a cafeteria, lounges, and game rooms. Other athletic facilities include the large athletic fieldhouse and the Navy Marine Memorial Stadium. Many of the buildings were erected with privately donated funds.

Each midshipman must satisfy certain minimum requirements in mathematics and science, the social sciences, and the humanities; at the end of the freshman year, he or she chooses a major, as the ongoing changes of technology make specialization more necessary than ever.

Entrance Requirements

1. Admission is open to young men and women of good moral character, without regard to race, creed, or national origin, but they must be citizens of the United States, must be at least seventeen years old but not past their twenty-second birthday on July 1 of the year of admission, and must be unmarried and have no children.

Candidates must (a) secure a nomination, (b) qualify scholastically, (c) meet prescribed physical standards, and (d) be selected for appointment.

Each member of Congress may have five appointees attending the Academy at one time, and each member may nominate up to ten candidates for each vacancy. Even if you are not selected to fill a Congressman's vacancy, if you have a nomination, a good school record, and meet the other standards, you have a chance as an alternate.

2. You must have acceptable scores on either the American College Testing Program Tests (ACT) or College Entrance Examination Board Scholastic Aptitude Test (SAT) and an acceptable school record including college preparatory work, in the upper 40 percent of the class.

3. You must pass the physical exam and the Physical Aptitude Test. It helps if you are conditioned by previous athletic participation.

4. You should apply in your junior year of high school and seek as many nominations as possible.

Evidence of leadership in extracurricular activities is helpful. The Admission Board makes selections on such factors as the above and comparative scores, and the number of women accepted is determined by the current needs of the Navy.

Complete information may be obtained by writing to:

Superintendent
U.S. Naval Academy
(Attn. Candidate Guidance)
Annapolis, MD 21402

You may want to visit your local Navy Recruiting Office, which can furnish pamphlets and sage advice.

(The Naval Academy Preparatory School at Newport, Rhode Island, now prepares enlisted men and women for entry into the Naval Academy.)

Questions and Answers

You may still have questions about the Naval Academy. The following are among those most often asked by prospective candidates. The answers may help to clear up any doubts or misunderstandings you may have.

Q. *What service selections are available to women?*
A. Federal law allows women to go to sea in ships that do not normally perform a combat mission (such as repair ships, research vessels, salvage vessels). Women aviators may also perform noncombatant flight duties, including the landing of aircraft aboard ships at sea. However, because of the limited number of noncombat-related sea billets available, only a limited number of women officers are able to select these billets.

Currently, the majority of women officer graduates develop other specialties and pursue careers ashore in one of many areas, including administration, communications, computer science, engineering, environmental science, legal, or research and development, to name just a few. It should be emphasized that most physically qualified women are commissioned in the line upon graduation from the Naval Academy.

A few women graduates, including those not physically qualified for a line commission, have the opportunity to be commissioned in such restricted line or staff communities as the Supply Corps, the Civil Engineer Corps, the intelligence community, etc. As is the case with men, opportunities available to women at graduation are subject to current needs of the Navy. All Marine Corps occupational specialties are available to women officers except infantry, artillery, tanks, and flying.

Q. *I don't know my Congressman. How do I get a nomination?*
A. It is *not* necessary to know him personally. Apply to the representative of your Congressional district and to both of your U.S. senators by mail; your applications will be considered carefully. Each member of Congress may have five appointees attending the Academy at any one time, and each member may nominate up to *ten* candidates for each vacancy. The essential thing to remember is that, by law, you *must* have a nomination to be considered for appointment. Once you are nominated, you officially become a candidate and your record can then be evaluated on its merits by the Naval Academy. Even if you are not selected to fill a particular Congressman's vacancy, if you have a Congressional nomination, have a good school record, and otherwise meet the basic entry standards, you will have an excellent chance to become a midshipman. Each year several hundred of the best qualified alternate Congressional nominees are appointed to the Naval Academy, as necessary, to bring the entering class up to authorized strength.

Q. *I'm in the Naval Reserve. Can I get into the Academy?*
A. Up to eighty-five reservists (Navy and Marine Corps), on active duty or members of drilling units, may qualify to enter the Academy each year through the reserves. See your career counselor for details.

Q. *If I am eligible for a Presidential nomination, should I also apply for a Congressional nomination?*
A. Yes. The more nominations you obtain, the better chance you will have for selection if you are found fully qualified.

Q. *Is it difficult to enter directly from high school?*
A. No. About seven out of ten midshipmen enter directly from high school. Of those who do not enter directly from high school, approximately 180 enter from the Naval Academy Preparatory School, 80 from the Naval Academy Foundation's preparatory scholarship program, 50 from private preparatory programs, 80 from colleges, and 20 from the Fleet (Navy & Marine Corps).

Q. *My grades were about average, but I played in several sports and was student body president. Also, I had to work after school. Will these activities help me?*
A. Yes. Evidence of leadership ability and participation in extracurricular activities, including athletics, part-time jobs, and civic activities, are considered in our evaluations.

Q. *Is physical preparation important?*
A. It certainly is! Aside from helping candidates to do well on the physical aptitude test, men and women alike should be as physically fit at entrance *as possible*, since the first summer is *very* demanding. Endurance and upper-body strength are particularly important. Cross-country runs, swimming, push-ups and chin-ups, and (particularly for women), the flexed-arm hang are valuable conditioning exercises. *Be in shape!* Or be sorry.

Q. *I have a high IQ and am a straight-A student. Will most of my time be spent on military subjects, or may I take any electives, such as electrical engineering?*
A. From 23 to 38 percent (depending on the major selected) of the Academy's curriculum is devoted to professional military studies, but at the Academy you will complete at least 140 rather than the 120 semester hours typical of most civilian colleges. Many academic majors are offered, from English to mathematics to physics to electrical engineering. Advanced research projects are offered in these and many other areas.

U.S. NAVY PHOTO

A seaman apprentice operates a crane in the dry dock at the submarine base at New London, Connecticut.

A. *What part of the medical examination gives the most difficulty to candidates?*
A. The eye examination. Visual acuity of 20/20 is required. However, a limited number of outstanding candidates may be granted waivers for visual acuity *if* the refractive error is not excessive and their vision is correctable to 20/20 with eyeglasses (*not* contact lenses). *No* waivers for defective color vision are granted. If within waiverable limits, and otherwise fully qualified for admission, you will be *automatically* considered for a waiver by the Academy's Admissions Board based on your overall record. Since only a limited number of nominees may be granted a medical waiver, the competition for the available waivers is keen. It should be noted that *all* nominees within waiverable limits (and otherwise fully qualified), including principal nominees of members of Congress, must compete for these waivers.

Q. *I don't like sports. Do I have to try out for anything?*
A. If you really dislike sports, then the Naval Academy may not be the best school for you. A midshipman is required to participate in athletics, either varsity or intramural, for the development of character, physical fitness, and competitive spirit.

Q. *How much does it cost to be a midshipman?*
A. Tuition, room and board, and medical and dental care are provided. In addition, midshipmen currently receive a monthly salary of $480 for uniforms, books, and personal needs. Salary and the value of the daily ration allowance ($3.80/day) accrue to a midshipman's pay account while on leave. A $1,000 deposit is required on entry. The deposit may be reduced or waived (upon request) in the case of serious hardship.

Q. *How often may I visit home?*
A. During Christmas and spring leaves. In addition, month-long summer leaves are granted to the three upper classes. You must pay for your own travel.

Q. *How many flunk out?*
A. About 8 percent of each entering class eventually leaves the Naval Academy because of academic failure.

Q. *Do I get to choose any of my courses?*
A. Yes, you will choose your major and the majority of your courses. The great majority of midshipmen get their first choice of a major. Occasionally, however, in order to better meet the future needs of the Navy, midshipmen must accept their second choice.

The Navy requires at least 80 percent of midshipmen to pursue engineering or science-related majors—a quota that we have had little trouble meeting voluntarily in recent years.

Q. *How much social life would I have at the Academy?*
A. Social life is *very* limited during the first year. After the initial (plebe) year, a growing range of social activities becomes available. In addition to weekend dances and other extracurricular activities at the Academy, there are opportunities for afternoon liberties in town and for a number of weekends away from the Academy.

Q. *I am a high school freshman. When should I start preparing myself for the Academy?*
A. Now! Your entire four-year high school record in academics and your record in athletics and other extracurricular activities for your last three years will be evaluated by the Naval Academy.

Q. *I am in college. Is it too late to enter the Academy?*
A. No, as long as you will not have passed your 22nd birthday on July 1 of the year of admission. Prior college work will permit study of advanced courses at the Academy. Normally, about 6 to 8 percent of the members of an entering class have been enrolled in a civilian college.

Q. *If I am not selected for an appointment for one class, am I eligible to apply for the next?*
A. Yes, as long as you still meet basic eligibility requirements pertaining to age, citizenship, etc., *and* obtain a new nomination. Also, each year, a number of the most promising of our unsuccessful civilian candidates are invited by the Academy to enlist in the Naval Reserve for the express purpose of attending the Naval Academy Prep School. Here, they become eligible to compete for entry into the Academy under a Secretary of the Navy nomination.

Q. *When should I apply for nomination?*
A. Apply to the representative from your Congressional district and to both your U.S. senators for a nomination, whenever possible, in the spring of your *junior* year in high school. Although a few apply as late as December of the senior year, this is not advised, since most members of Congress will have selected their nominees by that time.

Q. *What is my military obligation on graduation?*
A. Eight years. Current directives require five of these to be on active duty as a commissioned officer in the Navy or Marine Corps.

Q. *Does the Naval Academy have a pre-law major?*
A. The Academy has no pre-law program, and you might better plan to attend a civilian college offering pre-law studies if your primary interest is to become a Navy lawyer. Naval Academy graduates may apply for the Navy's law education program after two years of commissioned service. This is a *very* limited program, however, with only twenty-five officers selected from the entire Navy and Marine Corps each year. Officers not selected are required to continue their careers as line officers.

Q. *Can I take pre-medical studies at the Academy and become a doctor?*
A. An *extremely* limited number of midshipmen graduates may be selected to enter directly into medical school. No pre-med course of study is offered here. Prospective selectees must major in chemistry or other medically related science, and there is no guarantee of being selected into the program.

Q. *My father was in the armed forces. Will this help me to get a nomination?*
A. Sons and daughters of career members of the regular and reserve forces, active duty or retired (other than those retired under Section 1331 of Title 10, U.S. Code), may be considered for nomination under the Presidential category.

Q. *What if a midshipman is found to be smoking marijuana or to be using other unauthorized drugs?*
A. The Navy is *tough* on drugs. No person, officer or enlisted, is accepted into the Navy whose pattern of drug involvement indicates dependency, or who has a record of drug-trafficking offenses. All Induction Day physicals at the Naval Academy include a urinalysis testing for evidence of use of marijuana or other unauthorized drugs. This provides an informative data base and enables explicit counseling to be provided for those who need it. Illicit drug use while serving in the Navy is not tolerated. Failure to abide by this zero-tolerance drug policy—here or anywhere else—during your four years as a midshipman will result in separation from the Naval Academy. *Following* commissioning as an officer, such an offense could lead to trial by court martial, which can result in a punitive discharge and confinement at hard labor or an administrative discharge under other than honorable conditions.

Q. *Where do midshipmen live?*
A. They are housed in one large, multiwinged building, Bancroft Hall, having more than 4.8 miles of corridors and 33 acres of floor

U.S. NAVY PHOTO

A lieutenant (j.g.) mans equipment on a patrol hydrofoil during technical evaluation of the craft at a base in the Pacific Ocean.

space. Each room has its own shower, and all rooms have been remodeled in recent years. Fleet Admiral Ernest J. King Hall connects to Bancroft Hall. Here all 4,500 midshipmen are able to sit down and eat at one time. The food is served family style. Bancroft Hall contains a store, medical and dental facilities, a soda fountain, bowling alleys, and numerous other facilities.

Q. *I have nominations to both the Naval and the Air Force Academies. Must I undergo two medical examinations?*
A. No. A single medical examination conducted at any of the military examining centers designated by the Department of Defense Medical Examination Review Board is acceptable for all service academies.

Q. *What reasons are given most frequently by plebes who resign from the Academy?*
A. Resigning plebes most frequently say:
(1) They came to the Academy under parental pressure. After a few weeks as plebes, they feel that they have fulfilled their obligation to their parents and can safely resign.
(2) They were attracted to the Academy by its glamour. They knew that the academic program was demanding, but they failed to realize the extent of the daily demands made on their time by the military and professional aspects of the training at Annapolis. Some, apparently, were expecting more of a relaxed, college-type NROTC program than the regimen of a service academy.

Q. *How do I apply and what are my chances of getting into the Naval Academy Preparatory School?*
A. Whether you are military or civilian, no application is necessary. All nominees are considered *automatically* for entry into the Prep School by the Naval Academy's Admissions Board (based on their records) in the event they are unsuccessful in being selected for an appointment to the Naval Academy.

Q. *Is there a Naval Academy representative near my home town?*
A. There are more than 1,500 Naval Academy Information Officers available for counseling throughout the country.

Q. *Sea duty aboard a Navy ship sounds romantic, challenging, and all that, but how about family life? Family separations?*
A. Tough questions! A lot depends on you, your marriage, the way *your* family looks at the Navy and your career, the pluses and minuses of it, *after* you try it for awhile.

Most good marriages prosper in the Navy. Some may even be

made better... Navy families are a close, loyal, and proud group. Many feel their lives are enriched by the travel, friendships, and challenges offered through varied tours of duty in the Navy. But Navy life is certainly no cinch. Family separation *can* be a major problem, and it is frequently cited by those who elect to leave the Naval Service.

Navy Enlisted Careers

To enlist in the Navy men and women must be between seventeen and thirty-four years of age. Parental or guardian consent is required for seventeen-year-olds. You must be a U.S. citizen or an immigrant alien with permanent-residence status. A physical exam is required. While there are no specific educational requirements, the Navy prefers high school graduates. Both single and married men and women are accepted.

For young men and women who desire to learn and improve, the Navy offers real opportunity, including college training equivalent to junior college and practical experience.

Women are subject to the same regulations and are advanced and promoted like men. They have the same compensation, benefits, and privileges as men.

Navy Nurse Program

In the Navy Nurse Program, every area of practice is available: general nursing, clinical areas, and family practice such as pediatrics, obstetrics, and gynecology.

The Navy takes your choices into consideration, and you know your assignment before you join. Your first duty will probably be at a Navy hospital in South Carolina, Maryland, California, North Carolina, Texas, Florida, Washington, Tennessee, Connecticut, or Pennsylvania.

Graduate Navy nurses may also receive overseas assignments in such places as Naples (Italy), Bermuda, Puerto Rico, or Alaska, to name a few. Continuing education and higher degrees can be a part of a Navy nurse career.

The pay for a newly commissioned ensign, including allowances, is more than $17,000 a year. A Navy nurse with a direct commission as a lieutenant begins at more than $21,000 a year, including allowances.

Some of the financial benefits available to the Navy nurse are the initial $300 tax-free uniform allowance; thirty days' paid vacation each year; health and accident insurance; pension plan (retirement can begin after twenty years of service regardless of age); payment of costs for relocating; movies, golf, clubs, bowling, and swimming on base; and usually an option to live off-base and, if no Navy facilities are available, a housing allowance.

You wear your uniform on duty and for social occasions, but you can wear civilian clothes on your own time. And you'll look well in the uniform and enjoy wearing it. A uniform does something for a girl that no fancy dress could possibly achieve—and a nurse's uniform should make you very popular with the doctors and medical specialists; they will respect you as one of their own.

To sum up, the advantages of a Navy Nurse career include early responsibility in a variety of practice, chance for advanced education and possibly worldwide travel at Navy expense, career advancement, and unparalleled job security. The medical program officer at your nearest Navy recruiting station can give you further information.

Dental Student Program

To become a Navy Dental Student, you must be a citizen and be enrolled or have been accepted for enrollment in a dental school accredited by the American Dental Association. You must meet the personal qualifications and physical requirements for Navy commissioned officers.

You must be at least nineteen years old at the time of commissioning as an ensign, and under forty at the time of appointment as a lieutenant upon graduation from dental school. (The age limit may be adjusted upward by a maximum of 36 months for prior commissioned service.)

Naval Junior ROTC

Candidates for Naval Junior ROTC must be citizens of the United States, of good moral character, physically fit, and at least fourteen years old, and must attend a high school having an NJROTC unit.

The naval science curriculum is three years in length or may be four. A minimum of 180 class periods per year are supplemented by ship training cruises, orientation visits, and field trips to naval activities. In naval science, you learn seamanship, oceanography,

U.S. NAVY PHOTO

A Navy seaman is supervised during the process of plotting a weather front at the Naval Air Test and Evaluation Center, Patuxent, Maryland.

meteorology, astronomy, navigation, radar and sonar electronics, and have a chance for leadership. You may attend dances and military balls sponsored by your Junior ROTC unit, be a part of a color guard at school athletic events, and of course enjoy the field trips and tours of Navy ships and aircraft.

While you are under no obligation to join the armed services, there are certain bonuses if you do enlist. Naval JROTC can help you if you ever compete for an ROTC scholarship or an appointment to the Naval Academy or other service academies. Successful NJROTC completion also permits entry into the armed forces up to two pay grades higher than for those without JROTC background.

Navy ROTC Colleges and Universities

State	NROTC Unit	Crosstown Colleges/Universities
Alabama	Auburn University Auburn, AL 36830	
Arizona	University of Arizona Tucson, AZ 85721	
California	University of California at Berkeley Berkeley, CA 94720	*California State University—Hayward, Sacramento, San Jose *San Francisco State University *San Jose State University Stanford University University of California—Davis University of San Francisco *University of Santa Clara
	University of California at Los Angeles (UCLA) Los Angeles, CA 90024	California State University—Fullerton, Long Beach, Los Angeles, Northridge Loyola Marymount University Northrop University Occidental College Pepperdine University University of California—Irvine
	University of San Diego and San Diego State University San Diego, CA 92110 (Consortium)	University of California—San Diego
	University of Southern California Los Angeles, CA 90007	California Institute of Technology California State Polytechnic University Claremont McKenna College Harvey Mudd College
Colorado	University of Colorado Boulder, CO 80309	

State	NROTC Unit	Crosstown Colleges/Universities
Dist. of Columbia	George Washington University Washington, DC 20058	*American University *Catholic University of America *Georgetown University *Howard University *University of D.C. *University of Maryland
Florida	Florida A&M University Tallahassee, FL 32307	Florida State University Tallahassee Community College
	Jacksonville University Jacksonville, FL 32211	Edward Waters College Florida Junior College—Jacksonville University of North Florida
	University of Florida Gainesville, FL 32611	
Georgia	Georgia Institute of Technology (Georgia Tech) Atlanta, GA 30332	Agness Scott College Clark College Georgia State University Kennesaw College Morehouse College Morris Brown College Oglethorpe University Southern Technical Institute Spelman College
	Savannah State College Savannah, GA 31404	Armstrong State College
Idaho	University of Idaho Moscow, ID 83843	Washington State University
Illinois	Illinois Institute of Technology (Illinois Tech) Chicago, IL 60616	Elmhurst College Prairie State College Purdue University—Calumet *University of Chicago University of Illinois—Chicago Wilbur Wright City College
	Northwestern University Evanston, IL 60201	Loyola University
	University of Illinois Champaign, IL 61820	Parkland College
Indiana	Purdue University West Lafayette, IN 47907	
	University of Notre Dame Notre Dame, IN 46556	Bethel College Indiana University—South Bend St. Mary's College
Iowa	Iowa State University Ames, IA 50010	
Kansas	University of Kansas Lawrence, KS 66045	
Louisiana	Southern University A&M Baton Rouge, LA 70813	Louisiana State University
	Tulane University New Orleans, LA 70118	

State	NROTC Unit	Crosstown Colleges/Universities
Maine	**Maine Maritime Academy Castine, ME 04421	*University of Maine—Orono
Massachusetts	Boston University Boston, MA 02215	Northeastern University
	College of the Holy Cross Worcester, MA 01610	Anna Maria College Assumption College Central New England College Clark University Worcester Polytechnic Institute Worcester State College
	Massachusetts Institute of Technology (MIT) Cambridge, MA 02139	*Harvard University Tufts University *Wellesley College
Michigan	University of Michigan Ann Arbor, MI 48104	Eastern Michigan University—Ypsilanti
Minnesota	University of Minnesota Minneapolis, MN 55455	Augsburg College College of St. Thomas MacAlester College
Mississippi	University of Mississippi University, MS 38677	
Missouri	University of Missouri Columbia, MO 65201	Columbia College
Nebraska	University of Nebraska Lincoln, NB 68588	Nebraska Wesleyan University
New Mexico	University of New Mexico Albuquerque, NM 87131	University of Albuquerque
New York	Cornell University Ithaca, NY 14853	Ithaca College State University of New York—Cortland
	**New York State University Maritime College Fort Schuyler, Bronx, NY 10465	*Fordham University *Iona College *Manhattan College
	Rensselaer Polytechnic Institute Troy, NY 12180	Russell Sage College Siena College Skidmore College State University of New York—Albany Union College
	University of Rochester Rochester, NY 14627	Monroe Community College Nazareth College Rochester Institute of Technology St. John Fisher College State University of New York—Brockport State University of New York—Geneseo
North Carolina	Duke University Durham, NC 27706	North Carolina Central University
	University of North Carolina Chapel Hill, NC 27514	North Carolina State University

State	NROTC Unit	Crosstown Colleges/Universities
Ohio	Miami University Oxford, OH 45056	
	Ohio State University Columbus, OH 43210	Capital University Ohio Dominican College Otterbein College
Oklahoma	University of Oklahoma Norman, OK 73019	
Oregon	Oregon State University Corvallis, OR 97331	Linn Benton Community College
Pennsylvania	Pennsylvania State University University Park, PA 16802 (University Park campus only)	
	University of Pennsylvania Philadelphia, PA 19174	Drexel University LaSalle College St. Joseph's University Temple University
	Villanova University Villanova, PA 19085	Temple University
South Carolina	The Citadel Charleston, SC 29409 (Does not accept women)	
	University of South Carolina Columbia, SC 29208	
Tennessee	Memphis State University Memphis, TN 38111	
	Vanderbilt University Nashville, TN 37240	Belmont College David Lipscomb College Fisk University Tennessee State University Trevecca Nazarene College
Texas	Prairie View A&M University Prairie View, TX 77445	
	Rice University Houston, TX 77001	Houston Baptist University Texas Southern University University of Houston University of St. Thomas
	Texas A&M University College Station, TX 77843	
	Texas Tech University Lubbock, TX 79409	
	University of Texas Austin, TX 78712	
Utah	University of Utah Salt Lake City, UT 84112	Weber State College Westminster College
Vermont	Norwich University Northfield, VT 05663	

State	NROTC Unit	Crosstown Colleges/Universities
Virginia	Hampton Institute Hampton, VA 23668 Norfolk State University Norfolk, VA 23504 Old Dominion University Norfolk, VA 23508 (Consortium)	
	University of Virginia Charlottesville, VA 22903	
	Virginia Military Institute Lexington, VA 24450 (Does not accept women)	
	Virginia Polytechnic Institute and State University Blacksburg, VA 24061	
Washington	University of Washington Seattle, WA 98195	Seattle Pacific University Seattle University
Wisconsin	Marquette University Milwaukee, WI 53233	*Milwaukee School of Engineering Mt. Mary College University of Wisconsin—Milwaukee
	University of Wisconsin Madison, WI 53706	

*Crosstown enrollment is pending formal agreement between the schools. May not be available during school year 1986.

Navy Occupations

Career fields are listed alphabetically, beginning with those combat-related groups closed to women (indicated by "C").

Career Fields	Duties & Responsibilities	Qualifications	Examples of Civilian Jobs
Aviation Anti-submarine Warfare Operator (C)	Performs general flight crew duties; operates ASW sensor systems; performs diagnostic function to effect fault isolation and optimize system performance; operates tactical support center systems.	Above average learning ability. High degree of electrical and mechanical aptitude. Must pass flight physical and be able to swim. Courses in algebra, trigonometry, physics, electricity.	Radar technician, radio operator.
Electronic Warfare Technician (C)	Operates and maintains electronic countermeasures and electronic support measures; associated supporting equipment; evaluates, processes and applies intercept signal data, electronics intelligence reports, and electronic warfare tactics, and doctrine to operational needs.	Prolonged attention and mental alertness. Physics, mathematics and courses in radio and electricity helpful, and experience in radio repair. Normal sight and hearing, manual dexterity, and good memory.	Electronics technician, test equipment calibration technician.
Missile Technician (C)	Maintains fleet ballistic missiles and support equipment; tests, adjusts, calibrates, operates, repairs support equipment; handles/stows missiles.	High mechanical aptitude and manual dexterity, electricity, electronics, mathematics, and physics.	Rocket engine component ordnance artificer.

Career Fields	Duties & Responsibilities	Qualifications	Examples of Civilian Jobs
Molder	Operates foundries aboard ship and at shore stations; makes molds and cores, rigs flasks; casts ferrous, non-ferrous and alloy metals; sandblasts castings and pours bearings.	Foundry, machine shop, practical mathematics.	Foundry supervisor, furnace operator, melter, molder, core maker, heat treater, temperer.
Opticalman	Maintains binoculars, sextants, optical gunsights, turret and marine periscopes.	Close, exact and painstaking detail work, physics, shop mathematics and machine shop helpful; experience in optical or camera manufacturing.	Lens grinder, jewelry store cutter, tool inspector, optical maker, inspector, optical tooling specialist, camera repairer and locksmith.
Patternmaker	Makes wood, plaster, metal patterns, core boxes, flasks used by molders in Navy foundries.	Exacting, precise work; woodshop, foundry, mechanical awing, shop and practical mathematics.	Template maker, industrial arts teacher, layout person, patternmaker, form builder.
Aerographer's Mate	Collects, records, analyzes meteorological and oceanographic data; enters on appropriate charts; forecasts from visual and instrumental weather observations.	Algebra, geometry, trigonometry, physics, physiology, typing, training in meteorology and astronomy.	Weather observer, meteorological aide, chart maker, weather chart preparer.
Air Traffic Controller	Controls air traffic, operates radar air control ashore and afloat; uses radio, light signals; directs aircraft under visual flight and instrument conditions; assists in preparation of flight plans.	High degree of accuracy, precision, self-reliance and calmness under stress. Public speaking or experience in radio broadcasting.	Control tower operator, radio-telephone operator, airplane dispatch and aircraft log clerk.

Aircrew Survival Equipmentman	Maintains and packs parachutes, survival equipment, flight and protective clothing, life jackets; tests and services pressure suits.	Must perform extremely careful and accurate work. General shop and sewing desirable. Experience in use and repair of sewing machines.	Parachute packer, inspector, repairer and tester; sailmaker.
Aviation Anti-submarine Warfare Technician	Performs a wide range of electronic shop operations; performs in-flight maintenance of airborne electronic systems; removes and installs units of anti-submarine warfare equipment; debriefs flight crews; reads and applies service diagrams, schematics and manuals; maintains operating inventory of required equipment, tools and materials; uses and maintains a variety of test equipment.	Mathematical ability, manual dexterity, ability to do detailed work, a good memory, resourcefulness and curiosity.	Electronics mechanic.
Aviation Boatswain's Mate	Handles aircraft on carriers; operates, maintains, repairs aviation fueling, defueling, inert gas systems; maintains catapults, arresting gear.	20/20 vision uncorrected and good hearing. Shop work, physics and chemistry desirable. Experience in aircraft and hoisting equipment.	Machinery erector, crane operator, airport serviceperson, gasoline distributor. Firefighter-crash fire and rescue.
Aviation Electrician's Mate	Maintains, adjusts, repairs aircraft electrical and instrument systems, plus power generating, lighting, electrical components of aircraft controls.	Algebra, trigonometry, physics and shop experience in aircraft electrical work.	Aircraft electrician, electrician, and TV repairer.

Career Fields	Duties & Responsibilities	Qualifications	Examples of Civilian Jobs
Aviation Electronics Technician	Tests, maintains, repairs aviation electronics equipment including navigation, identification, detection, reconnaissance, special purpose equipment; operates warfare equipment.	High degree of aptitude for electrical and mechanical work. Algebra, trigonometry, physics, electricity, radio and mechanics.	Aircraft electrician, radio mechanic, electrical repairer, instrument repairer, electronics technician, TV repairer.
Aviation Fire Control Technician	Maintains and inspects aircraft weapons systems, weapon-control radar, computers, computer sights, gyroscopes, related equipment; air launched guided missile equipment.	Superior electronic, electrical and mechanical aptitude. Training in repair shops or vocational schools and in mathematics.	Instrument person, aircraft electrician, electronics technician, radar computer repairer, TV repairer.
Aviation Machinist's Mate	Inspects, maintains power plants and related systems and equipment, prepares aircraft for flight, conducts periodic aircraft inspections.	Good learning ability and mechanical aptitude. Machine shop, automobile or aircraft engine work, algebra and geometry.	Airport serviceperson, aircraft engine test mechanic, small appliance repairer.
Aviation Maintenance Administration	Management and clerical duties in aircraft maintenance offices, plans and schedules maintenance workload, prepares reports and correspondence and analyzes trends of aircraft system and component failures.	Accurate and detailed work, has interest in the aviation maintenance field. Filing and typing.	Shipping, parts, supply room or maintenance clerk, office manager.
Aviation Ordnance	Loads bombs, torpedoes, rockets, guided missiles; maintains, repairs, inspects aircraft armament, aviation ordnance equipment.	Normal vision and good mechanical aptitude. Algebra, physics and electricity experience in electrical or mechanical repair.	Gyroscope mechanic, instrument person, ordnanceman, armament inspector.

Aviation Storekeeper	Receives, stores, issues aviation supplies, spare parts, technical aviation items; conducts inventories.	Bookkeeping, accounting, business math and typing helpful.	Clerk typist, inventory, material, sales, or receiving/shipping.
Aviation Structural Mechanic	Maintains and repairs aircraft, airframe, structural components, hydraulic controls, utility systems, egress systems.	High degree of mechanical aptitude. Metal shop, wood-working, algebra, plane geometry, physics; experience in automobile body work.	Welder, sheet metal repairer, hydraulics technician, air conditioning repairer.
Aviation Support Equipment Technician	Services, tests and performs intermediate and repair of gasoline and diesel engines, gas turbine compressor units, power generating equipment, liquid and gaseous oxygen and nitrogen servicing equipment, automotive electrical and air conditioning systems.	High mechanical aptitude. Mathematics, physics, electricity and machine shop and experience as auto mechanic or machinist.	Diesel or gasoline engine, air conditioning ignition mechanic, hydraulics repairer, auto mechanic.
Boatswain's Mate	Performs seamanship tasks, operates small boats, stores cargo, handles ropes and lines, directs work of deck force personnel.	Must be physically strong. Practical math desirable; algebra, geometry and physics.	Motorboat operator, pier superintendent, able seaman, canvas worker, rigger, cargo wincher, mate, longshoreperson.
Boiler Technician (C)	Operates boilers and fireroom machinery; transfers, tests and takes inventories of fuel and water, maintains boilers, pumps, associated machinery.	Strong interest in mechanical work. Shop courses and practical mathematics.	Marine fireman, boiler ship mechanic, boiler maker, stationary engineer, boiler or heating plant.

Career Fields	Duties & Responsibilities	Qualifications	Examples of Civilian Jobs
Builder	Constructs, maintains, repairs wood, concrete, masonry structures; erects and repairs waterfront structures.	High mechanical aptitude. Carpentry and ship mathematics desirable. Experience with hand and power tools valuable.	Plasterer, roofer, mason, painter, construction worker, carpenter.
Construction Electrician	Installs, operates, maintains, repairs electrical generating and distribution systems, transformers, switchboards, motors, controllers.	Interest in mechanical and electrical work. Electricity, shop mathematics and physics helpful, ability to work aloft.	Powerhouse or construction electrician, electrical and telephone repairer.
Construction Mechanic	Maintains, repairs and overhauls automotive and heavy construction equipment.	High mechanical aptitude. Electrical or machine shop, ship mechanics and physics helpful. Machinist or auto mechanic work.	Automotive or diesel engine mechanic, motor analyst, construction equipment mechanic.
Cryptologic Technician	All prospective Cryptologic Technicians (CTs) must have a personal background that will facilitate clearance for special security access; above average speaking and writing ability; good memory; resourcefulness; curiosity; adaptability to detailed work; aptitude for math; record-keeping ability; ability to work well with others and manual dexterity.		
Cryptologic Technician A (Administrative)	Types messages and correspondence; files; handles classified material; keeps mail logs; prepares correspondence; orders supplies and takes inventory.	Exceptionally good character, speaking and writing ability.	Clerk typist, office manager.
Cryptologic Technician I (Interpretive)	Operates technical communications systems equipment; prepares data and reports involving communications material; performs temporary duty aboard submarines and surface units.	Foreign language aptitude.	Interpreter, translator.

Cryptologic Technician M (Maintenance)	Performs preventive and corrective maintenance on solid-state and electro-mechanical equipment which requires use of test equipment, hand tools and technical publications; repairs and calibrates wide variety of precision electronic test equipment.	Ability to comprehend advanced electronic theory.	Automatic equipment technician; central office repairer, radio mechanic.
Cryptologic Technician O (Communications)	Prepares messages utilizing teletypewriter equipment; transmits, receives, routes and logs message traffic; maintains message center files/logs and records; controls and operates communications equipment systems including radio receivers, tone-terminal equipment, DC and audio patch boards and communication security devices.	Typing ability, good memory, resourcefulness, curiosity, manual dexterity, aptitude for figures.	Cryptographic machines, telegraphic-teletypewriter operator.
Cryptologic Technician R (Collection)	Variety of duties associated with operation of teletype and Morse communications systems; operates radio-receiving, direction-finding; and technical documents which are predominantly classified.	Good memory, speaking and writing ability.	Morse operator; radio officer.

Career Fields	Duties & Responsibilities	Qualifications	Examples of Civilian Jobs
Cryptologic Technician T (Technical)	Variety of duties associated with the operation of radio printer and other sophisticated equipment to study signal propagation; operates radio receiving, teletype, recording and related computer equipment; Morse code, security, communications procedures.	Above average speaking and writing ability, good memory.	Cryptanalyst, digital computer operator; electronic intelligence operation specialist and telegraphic-typewriter.
Data Processing Technician	Operates data processing equipment including sorters, collators, reproducers, tabulating printers and computers.	High clerical aptitude. Typing, bookkeeping and operating business machines desirable. Experience in mechanical work.	Data typist, key punch operator, systems analyst, verifier and tabulating machine operator.
Data Systems Technician	Maintains electronic digital data systems and equipment; inspects, tests, calibrates, and repairs computers, tape units, digital display equipment, data link terminal sets and related equipment.	Possess high aptitude for detailed mechanical work. Radio, electricity, physics, and mathematics through calculus.	Electrical, electronic repairer or computer repairer, electronic and data systems technician.
Dental Technician	Assists dental officers in treatment of patients; performs preventive procedures, and various dental department administrative duties.	Hygiene, physiological and chemistry.	Dental technician, dental hygienist, X-ray technician, dentist's assistant, dental laboratory technician.
Disbursing Clerk	Opens, maintains, closes military pay records; prepares reports and returns on public monies.	Typing, bookkeeping, accounting, business math and office practices.	Paymaster, cashier, statistical or audit clerk, bookkeeper, bookkeeping machine operator, cost accountant.

Electrician's Mate	Maintains power and lighting equipment generators, motors, power distribution systems, other electrical equipment; rebuilds electrical equipment.	Aptitude for electrical and mechanical work. Electrical, practical and shop mathematics, and physics.	Electrician, electric motor and electrical equipment repairer, armature winder, radio/TV repairer.
Electronics Technician	Maintains all electronic equipment used for communications, detection ranging, recognition and countermeasures.	Aptitude for detailed mechanical work. Radio, electricity, physics, algebra, trigonometry and shop valuable.	Electronics technician, radar and radio repairer, instrument and electronics mechanic.
Engineering Aid	Performs tasks required in construction surveying in drafting. Makes and controls surveys, runs and closes traverses; conducts soil classification and compaction tests.	Algebra, geometry, trigonometry, mechanical drawing and drafting recommended. Experience in road construction useful.	Surveyor, draftsperson, soil analyst.
Engineman	Operates, maintains, repairs internal combustion engines, main propulsion machinery, refrigeration and assigned auxiliary equipment.	Algebra, geometry and physics helpful. Experience in automotive repair.	Diesel engine operator, diesel mechanic, ignition repairer, small engine mechanic, marine oiler, stationary engineer.
Equipment Operator	Operates automotive and heavy construction equipment.	Good physical strength and normal color perception. Experience in construction work, auto or electrical shop.	Bulldozer, pile driver, power shovel or motor grader operator; excavation foreman, truck driver.
Fire Control Technician (C)	Operate, test, maintain and repair weapons control systems and telemetering equipment used to compute and resolve factors which influence accuracy of naval guns and missiles.	Perform fine, detailed work. Extensive training in mathematics, electronics, electricity and mechanics.	Radar or electronics technician, test range tracker, instrument repairer, electrician.

Career Fields	Duties & Responsibilities	Qualifications	Examples of Civilian Jobs
Gas Turbine Systems Technician (C)	Operates gas turbine engines, main propulsion machinery and related electrical and electronic equipment.	Electrical/electronics repair, blueprint reading, mathematics and physics.	Electronics technician, gas turbine mechanic, power plant operator.
Gunner's Mate (C)	Operate and perform maintenance on guided-missile launching systems, rocket launchers, guns, gun mounts; inspects/repairs electrical, electronic, pneumatic, mechanical and hydraulic systems.	Prolonged attention and mental alertness, ability to perform detailed work. High aptitude for electrical and mechanical work. Arithmetic, shop math, electricity, electronics, physics, machine shop, shopwork, welding, mechanical drawing.	Electronics technician, electrician, instrument repairer, hydraulics, pneumatic or mechanical technician, small appliance or test equipment repairer.
Hospital Corpsman	Administers medicines, applies first aid, assists in operating room, nurses sick and injured.	Hygiene, biology, first aid, physiology, chemistry, typing and public speaking.	Practical nurse, medical or X-ray lab technician, nurse administrator.
Hull Maintenance Technician	Fabricates, installs, repairs shipboard structures, plumbing and piping systems; uses damage control in firefighting, and nuclear, biological, chemical and radiological defense equipment.	High mechanical aptitude. Sheet metal, foundry, pipefitting, carpentry, mathematics, geometry and chemistry valuable.	Firefighter, welder, plumber, shipfitter, blacksmith, metallurgical technician.
Illustrator Draftsman	Designs, sketches, does layout, letters signs, charts and training aids; operates visual presentation equipment; makes mathematical computations for layout and design illustrations.	Previous experience as drafting person, tracer or surveyor valuable. Art, mechanical drawing and blueprint reading valuable.	Structural drafting person, technical illustrator, specification writer, electrical drafting person, geodetic computer, graphic artist.
Intelligence Specialist	Maintains/uses intelligence files; prepares maps, graphics, mosaics, charts; extracts intelligence information from aerial photos;	Processing, assimilating, interpreting and presenting data. Typing, filing, drafting, mathematics, geography and	Photographer, aerial picture analyst, chart maker, navigation instructor.

Interior Communications Electrician	Maintains, operates all interior communications systems, voice interior communications, alarms, ship's control, plotting, automated propulsion equipment.	High aptitude for electrical work. Electrical shop, practical shop mathematics, experience in electrical/electronics work desirable.	Powerhouse engineer, ship electrician, station installer, instrument person, electronics or TV technician.
Journalist	Reports, edits, copyreads news; publishes information about service people and activities through newspapers, magazines, radio and television.	High degree of clerical aptitude. English, journalism, typing and writing experience helpful.	News editor, copyreader, script writer, reporter, free lance writer, rewrite or art layout person, producer.
Legalman	Provides administrative services, military justice, claims, administrative law, and legal assistance; serves as court reporter.	Aptitude for detail, ability to express self in writing and orally. Typing, shorthand, English and logic helpful. No speech or hearing difficulties.	Legal assistant, law and contract clerk, title examiner and court reporter, office manager.
Lithographer	Performs offset lithography and letterpress printing, copy preparation, camera work, assembling and stripping, platemaking, typesetting, presswork and binding.	Work with machinery and chemicals. Printing, physics, chemistry, English and shop mathematics valuable.	Lithographic and plate press operator, bookbinder, printer, photoengraver, cameraperson, and photolithographer.
Machinery Repairer	Maintains assigned equipment to support other ships requiring use of milling machines, boring mills, other machine tools found in machine shop, overhaul and repair machinery.	Experience in practical or shop mathematics, machine shop, electricity, mechanical drawing and foundry desirable.	Engine lathe operator, machinist tool clerk, bench machinist, turret and milling machine operator and tool maker.

Career Fields	Duties & Responsibilities	Qualifications	Examples of Civilian Jobs
Machinist's Mate (C)	Operates, maintains and repairs ship's propulsion, auxiliary equipment and outside equipment such as steering, engine, refrigeration/air conditioning, laundry equipment.	Aptitude for mechanical work. Practical or shop mathematics, machine shop, electricity and physics valuable.	Boiler house repairer, engine maintenance machinist, marine engineer, turbine operator, engine repairer, air conditioning and refrigeration repairer.
Master-at-Arms	Performs investigations, apprehensions, crime prevention, preservation of evidence; performs duties of beach guard and shore patrol, crowd control and brig operations.	Experience in police, shore patrol or investigative work. Maturity, good vision and hearing. High school diploma or equivalent.	Police officer, guard, detective, investigator.
Mess Management Specialist	Operates and manages Navy dining facilities and bachelor quarters; estimates quantities and kinds of foodstuffs required; receives, stows and breaks out food items; prepares menus; plans, prepares and serves meals; maintains stock records; conducts inventories; assists medical personnel in inspection for quality; complies with sanitary and hygienic requirements.	Experience or courses in food preparation, dietetics and record keeping helpful. High standards of honesty and cleanliness; good learning ability.	Caterer, cook, steward, chef, mess attendant, motel/hotel services assistant.
Mineman	Tests, maintains and repairs mines, components and mine laying equipment.	High mechanical aptitude. Electricity, machine shop work, mechanical drawing and shop mathematics desirable.	Ordnanceman, mine assembler, ammunition foreman, powderman, electrician, harbor patrol.

Musician	Provides music for military ceremonies, religious services, concerts, parades, various recreational activities; plays one or more musical instruments.	Proficiency on standard band or orchestral instruments.	Music teacher, instrument musician, orchestra leader, music arranger, instrument repairer, music librarian, arranger.
Ocean Systems Technician	Operates special electronic equipment to interpret and document oceanographic data; operates related equipment such as tape recorders; interprets data; prepares and maintains visual displays of data; converts data into formats for statistical study.	Normal hearing, vision and color perception; above average learning ability; ability to perform detailed and repetitive work, to work harmoniously with others, and with numbers; and qualified for secret security clearance.	Computer-peripheral equipment operator or electronic technician.
Operations Specialist	Operates surveillance and search radar, electronic recognition and identification equipment, controlled approach devices and electronic aids to navigation; serves as plotter and status board keeper.	Prolonged attention and mental alertness. Physics, mathematics and ship courses in radio and electricity helpful. Experience in radio repair is valuable.	Radio operator (aircraft, ship government service, radio broadcasting), radar equipment supervisor and control tower operator.
Personnelman	Performs enlisted administration duties in manpower utilization, maintains service records, personnel accounting, educational services, classifies personnel and jobs.	Ability to deal with people, typing, public speaking, office practices, personnel work and counseling helpful.	Employment manager, personnel clerk, vocational adviser, clerk typist, job or organizational analyst.
Photographer's Mate	Operates, maintains and repairs cameras for ground and aerial photographic work.	Normal color perception; physics and chemistry desirable.	Photographer, camera repairer, aerial photographer.

Career Fields	Duties & Responsibilities	Qualifications	Examples of Civilian Jobs
Postal Clerk	Processes mail, sells stamps and money orders, maintains mail directories and handles correspondence concerning postal operations.	Bookkeeping, accounting, business math, typing and office practices.	Parcel post or mail clerk, mail room manager, stock clerk, cashier.
Quartermaster	Performs navigation of ships, steering, lookout supervision, ship-control, bridge watch duties, visual communication and maintenance of navigational aids.	Good vision and hearing and ability to express oneself clearly in writing and speaking. Public speaking, grammar, geometry and physics helpful.	Barge, motorboat, yacht captain, quartermaster, harbor pilot aboard merchant ships.
Radioman	Operates communication, transmission, reception, and terminal equipment; transmits, receives and processes all forms of military record and voice communications.	Good hearing and manual dexterity. Mathematics, physics and electricity desirable. Experience as amateur radio operator helpful.	Telegrapher, radio dispatcher, radio/telephone operator, news copywriter.
Religious Program Specialist	Assists in management of religious programs and facilities; trains volunteers; supervises the offices of chaplains; performs administrative duties.	Relate easily with people. Basic English, business math, typing, graphics and audiovisual familiarization helpful.	Church business manager or administrator.
Ship's Serviceman	Operates and manages ship's store activities afloat and ashore, including barber, cobbler, tailor, laundry, dry cleaning, commissaries, retail stores.	Shoe repairing, barbering, tailoring, merchandising and retailer, accounting, bookkeeping, business math and English helpful.	Barber, laundry and dry cleaner, retail store manager, sales clerk, tailor, and shoe repairer.

Sonar Technician (C)	Operates electronic underwater detection and attack apparatus, obtains and interprets information for technical purposes, maintains and repairs electronic underwater sound detection equipment.	Normal hearing and clear speaking voice. Algebra, geometry, physics, electricity and shopwork desirable. Experience as amateur radio operator.	Oil well sounding device operator, radio operator, inspector of electronic assemblies, electronic technician, electrical repairer.
Signalman	Sends and receives messages by flashing light, semaphore and flag hoist; handles, routes and files messages; codes and decodes message headings; operates voice radio and maintains visual sight equipment.	Good vision and hearing, ability to express oneself clearly in writing and speaking.	Third mate, signal officer, deck cadet, harbor police officer, small boat.
Steelworker	Fabricates, erects and dismantles pre-engineered structures, steel bridges and other structures. Lays out and fabricates steel and sheet metal; welds.	Physical strength, stamina and ability to work aloft. Sheet metal, machine shop, foundry experience desirable.	Rigger, shipfitter, structural steelworker, salvage engineer, steel fabricator, welder, sheet metal technician.
Storekeeper	Orders, receives, stores, inventories and issues clothing, foodstuffs, mechanical equipment and other items. In the Coast Guard also performs duties as disbursing clerk.	Typing, bookkeeping, accounting, commercial math, general business studies and English helpful.	Sales or shipping clerk, warehouse worker, buyer, invoice control clerk, purchasing agent.
Torpedoeman's Mate	Maintains and overhauls torpedoes and depth charges; maintains and repairs ordnance launching equipment; launches and recovers torpedoes.	High mechanical and electrical aptitude. Electricity, machine shop, welding, mechanical drawing and ship mathematics desirable.	Ordnance supervisor, gyroscope assembly supervisor, instrument mechanic, electronics technician, motor/office machine repairer.

Career Fields	Duties & Responsibilities	Qualifications	Examples of Civilian Jobs
Utilitiesman	Installs, maintains, repairs and codes plumbing, heating systems, steam, compressed air, fuel storage, collection and disposal facilities and water purification units.	High mechanical aptitude. Apprentice training in plumbing and related fields, mathematics helpful.	Stationary engineering assistant, plumber, pipe fitter, water plant or boiler operator, boiler house supervisor.
Yeoman	Clerical and secretarial, typing, filing, operating office and duplicating equipment, preparing and routing correspondence and reports, maintains records and official publications.	Same qualifications required of secretaries and typists in private industry; English, business subjects, stenography and typewriting helpful.	Office manager, secretary, general office clerk, administrative assistant.

The Marine Corps

The Marine Corps has its own Commandant, who is responsible to the Secretary of the Navy and gives technical advice to the Chief of Naval Operations. The Commandant also is responsible for material and training to prepare Marines for combat duty. Some Marines go to sea; others are stationed at shore establishments of the Navy as security forces.

In World War II, Marines used pontoon landing gear, platform-like equipment. Since then helicopters have come into use to airlift combat troops from offshore ships for amphibious landings. Later helicopters bring equipment and supplies when the Marines have established a shore base.

During World War II, legislation based on the needs of the nation enabled women to enter the United States Marine Corps. Later laws ensured their continuing service, and current laws provide them with equality of opportunity. Currently some 9,666 officer and enlisted women Marines are on active duty and over 1,800 are in the drilling Reserve components—more than at any other period except World War II when the Marine Corps rose to an all-time high of 485,000 Marines, including about 20,000 women Marine reservists.

Qualifications for Commission or Enlistment

Officers. To qualify for acceptance as a woman Marine officer candidate, the following requirements are necessary:

1. American citizenship (aliens are eligible under certain conditions).
2. A baccalaureate degree or enrollment as a college junior or senior, carrying a full daytime academic program leading to a baccalaureate degree, and maintaining a C average or higher.
3. Minimum age of twenty, and less than twenty-eight at the time of commissioning. Waivers may be granted down to age nineteen or up to age thirty.
4. The mental, moral, and physical standards determined by the Marine Corps as essential for a commission as a Marine officer.

Enlisted. Minimum qualifications for the enlistment of young women in the Marine Corps are:

1. American citizenship (aliens are eligible under certain conditions).
2. Minimum age of seventeen, maximum of twenty-eight.
3. High school graduate or equivalent education. For example, a nonhigh school graduate who has completed at least thirty semester

hours with a C+ average or better at an accredited college may be accepted for enlistment. Prior to October 1, 1978, a General Education Diploma (GED) was considered qualifying for a specified number of women seeking to enter the Marine Corps; however, it is no longer considered sufficiently qualifying for enlistment.

4. Excellent health and moral character.

With college tuition costs increasing an average of 7 percent, soaring in some cases to $17,210 for one year, and increasing at even relatively inexpensive private and public colleges (by 5 percent even in two- and four-year public colleges), the quality education offered by the military is a good bet for students interested in a military career.

The training at Parris Island, S.C., is designed to produce a basic woman Marine. Recruits become sufficiently familiar with individual weapons and field skills to meet the minimum training requirements of the Marine Corps woman Marine assignment policy. The overall goal of recruit training is to instill in the recruit skills, knowledge, discipline, pride, and self-confidence so that at graduation she will be worthy of recognition as a United States Marine.

Recruits drill in platoons and learn to perform as a team. The

U.S. MARINE CORPS PHOTO

An instructor supervises calisthenics during a training session at the Marine Corps Recruit Depot, Parris Island, South Carolina.

intensive 11-week program is divided into three phases, all of which are professionally administered and supervised. In the first phase emphasis is on physical conditioning, close-order drill, first aid, Marine Corps history and tradition, military customs and courtesies, the Uniform Code of Military Justice, Interior guard, uniform regulations, and water safety survival. Appearance, weight in proportion to height, and good grooming are emphasized along with responsiveness and performance. An hour of free time is provided at the end of the day and more free time on Sunday. Religious services are also available. The many hours spent on physical fitness are designed to enhance the recruit's appearance and self-esteem and provide the increased stamina and muscle tone required. Depending on the program, most days begin at 5:30 a.m. and end around 9:30 p.m. Saturdays and Sundays are usually lighter in schedule.

In the second training phase, the recruits attend classes on nuclear, biological, and chemical defense, field sanitation, preventive dentistry, and hygienic standards. They also continue the rifle familiarization classes, and spend a half day on the rifle range, where they fire the M-16. Safety in handling and firing the weapon is stressed. Recruits also participate in field training, receiving instruction in field living and orientation in basic defensive tactics.

During the third and final phase of training, the recruits are instructed in a variety of subjects, including personal and financial responsibilities, occupational specialty training in formal schools available to women in the Marine Corps; career opportunities, and the Corps' policy on drug and alcohol abuse.

After final exams, the physical fitness test, and the intensive Command Inspection comes the day before graduation, when the recruit wears the Marine Corps emblem for the first time. Upon graduation, the young woman is addressed as a Marine for the first time. Graduation ceremonies, or Final Review, are usually held at the Recruit Depot's main parade deck, and family and friends are invited to attend.

The new Marine is now ready to begin Military Occupational Specialty (MOS) training at one of the fourteen occupational fields open to Women Marines.

Title 10 of the United States Code, Section 6015, prohibits women serving in the Department of the Navy from assignment "...to duty on vessels or in aircraft that are engaged in combat missions, nor may they be assigned to other than temporary duty on vessels of the Navy, except hospital ships, transports and vessels of a similar classification not expected to be assigned combat missions."

Following these guidelines, the Marine policy states that women

Marines may be classified in any noncombatant occupational field for which they are qualified. The four fields in which women may not serve are infantry, field artillery, tank and amphibian tractor, and pilot and naval flight officer.

Opportunities for women continue to open up in more and more occupational specialties, duty stations, and training. In 1988 the first women in nine years were graduated from Marine Security Guard school. They underwent extensive training—the same as their male counterparts—and they were placed in challenging assignments as Marine Security Guards.

So successful have women Marines been that a recent review called for increasing enlisted women Marine strength to 10,400—up from 8,500, a gradual increase reaching its final objective in 1990.

The Marine Corps also needs women officers, and leadership is a tenet that has been central to the Marine Corps for over 200 years. True leadership is an art. It is the influencing and direction of people with resulting obedience, respect, confidence, and loyal cooperation.

No one is born a leader. No classroom or graduation certificate confers leadership. It evolves in the crucible of necessity, formed from the basic elements of observation and experience. The Corps does not see the relation between officers and enlisted personnel as that of superior and inferior or of master and servant, but rather that of leader and scholar. In other words, the role of the officer goes beyond providing discipline and military training to a concern for the welfare of the young persons in service.

The Marine Corps can offer a young man or woman physical training, academic advancement, meaningful employment, economic security, and a variety of interesting experiences and travel. The leadership and dedication are up to you.

The first training ground for the majority of Marine Corps officers is college. Commissioning programs require at least a bachelor's degree. The official Marine Corps policy is "Stay in school." The Marine Corps Officer Selection Officer who visits your campus can provide all the up-to-the minute details.

For male freshmen, sophomores, and juniors, and female juniors, Platoon Leaders Class (PLC), takes place in the summer at Officer Candidate School, Marine Corps Combat Development Command, Quantico, Virginia. Travel costs to and from Quantico, meals, textbooks, etc., are furnished by the government.

You may also receive $100 per month for a nine-month school year in exchange for additional active-duty obligations. This financial assistance is payable for up to three years, or a total of $2,700. You must maintain an overall C average, and upon graduation you are commissioned a Marine second lieutenant.

PLC-Law is a post-baccalaureate degree program to meet the Marine Corps need for law specialists. In this program active duty is postponed until you have obtained a law degree and been admitted to the bar. Meanwhile you're commissioned and promoted on schedule.

Marine Option NROTC

There are two ways to take advantage of this program. First is the Scholarship program, for college-bound high school students. If you're selected you attend one of more than sixty-five colleges and universities across the nation and receive substantial financial assistance. During your college training, you receive tuition, cost of textbooks, fees, and a subsistence allowance of $100 per month for a maximum of forty months.

Military education is enhanced with a six-week summer training cruise aboard a U.S. Navy ship or a stint at a major shore installation following your freshman year. The summer after your sophomore year includes four weeks of warfare specialties training. During the summer after your junior year, you attend a precommission training period with the Marine Corps.

These positions are limited and go only to the most highly qualified candidates. You must apply before or early in your senior year of high school. See your graduate counselor or Marine recruiter for complete details.

The second NROTC Marine Option is the College Program, for students already in college. It provides selectees $100 per month during their junior and senior years, but no scholarship benefits. One precommission training period of ten weeks is the only summer session required for women. The Women Officer Candidate Program (WOCP) for juniors, seniors, and college graduates offers this ten-week summer training following the junior year or after graduation.

In consolidated Officer Candidate companies at Quantico, you participate in much of the same rigorous screening programs as your male counterparts. Women who complete the summer training session following their junior year are eligible for financial assistance as seniors.

In most occupational fields, you work side by side and on a parity with men. You have the chance to do almost any job in the Marine Corps with the exception, as explained earlier, of those that might put you in a combat situation.

The WOCP also offers a law option like that of PLC (Platoon Leaders Class) and OCC (Officer Candidate Class).

Whether you enter PLC, OCC, NROTC, or WOCP, you undergo training at the Officer Candidate School, Quantico, Virginia. It's many weeks of some of the toughest physical training in the world. It's demanding. You stop counting the number of pull-ups, push-ups, sit-ups you do and miles you run. You learn to negotiate an obstacle course you never dreamed you could handle, and you do it faster every time. You learn to take commands and to give them. You learn your rifles inside and out. The emphasis on leadership remains.

You are given temporary leadership positions so that the instructors can evaluate your appearance, your speech, your command presence, your strength, agility, coordination, and endurance, your intelligence, and your moral and physical courage. But above all else you are evaluated on your ability to lead other Marines under conditions of extreme stress.

About halfway through The Basic School, you make a major decision: what kind of occupation you'd like to pursue as an officer of the Marines. You are asked to list a first, second, and third preference of occupational fields. Whether or not you are assigned to your preference depends on four factors: your preference, your class standing, your company commander's recommendation, and Marine Corps requirements. Your chances are excellent, however. Approximately 90 percent of all graduates from The Basic School are assigned to the occupational field of their choice.

And your choice is not limiting. During your career, you will work not only in your occupational field, but also in a wide variety of other assignments.

Some of the career options are: aviation, special support, air control, aircraft maintenance, legal, data processing, supply, and communications. Continuing education is offered at many U.S. schools as well as Canadian Forces Staff College, Toronto, Ontario, Canada; Spanish Naval War College, Madrid, Spain; and Staff and Command College of the Federal German Armed Forces, Hamburg, Germany. At the apex of your career, in addition to U.S. special schools, opportunities include British Imperial College, London, England; NATO Defense College, Rome, Italy; Norwegian National Defense College, Oslo, Norway; Australian Joint Service

Staff College, Canberra, Australia; and National Defense College of Canada, Kingston, Ontario, Canada.

Under the Marine Graduate Education program, you may have the opportunity to pursue an advanced degree at a civilian college full-time, at Marine Corps expense, to perfect your skills in such fields as aeronautical engineering, statistics, communications, engineering, computer science, and defense systems analysis.

The woman Marine's expertise will not be just physical but mental. Every base has a library, and if you like to read you'll collect your own books. Technology does not remain the same; nor will the Marine Corps. Even now, there is a trend away from use of firepower tactics toward a maneuverability tactic that requires keen mental insights to "outfox" an enemy rather than killing him. In the world of international affairs, global thinkers will make the future.

Marine Corps Education

The Marine Corps maintains more than 500 schools, both basic and advanced, to provide formal training to enlisted Marines. These are geared into occupational needs. See the following Occupation List.

The Marine Corps Institute is an official training activity of the Marine Corps that prepares and administers correspondence courses. These are provided free to Regular and Reserve Marines of any rank. Training is available in such fields as electronics, food service, and law enforcement. A naval justice course trains nonlawyer personnel for legal duties. Other specialty courses include civil engineering, transportation, petroleum products, and supply.

Correspondence courses are available from the Naval Correspondence Course Center, the Naval War College, the Industrial College of the Armed Forces, the U.S. Medical and Dental School, the Defense Intelligence School, and the Submarine School. Naval officers may also take courses offered by the Army and Air Force or enroll in graduate or undergraduate programs in colleges and universities.

For instance, the Marine Corps Enlisted Commissioning Education Program (MECEP) enables selected enlisted Marines to earn a bachelor's degree by attending a college or university as a full-time student. Those who earn a degree and complete officer candidate training are commissioned second lieutenants in the regular Marine Corps. The program has scholastic and age requirements and a service obligation upon graduation.

Officer Candidate Schools provide a way for selected college graduates to earn commissions as line officers, Supply Corps or Civil Engineer officers, in intelligence, as oceanographers, meteorologists, or naval security specialists. To become a naval aviator, you must be a graduate of an accredited college or university. The Aviation Officer Candidate School is at Pensacola, Florida. Aviation Maintenance instruction for college seniors or graduates is at Millington, Tennessee, and Athens, Georgia.

To enter a Marine Women Officer Candidate Class following your junior year of college (or after graduation), you should plan ahead. The Officer Candidate school is at Quantico Marine Corps Base, Quantico, Virginia. Candidates must be a minimum of eighteen years of age when enrolled or between twenty and thirty when commissioned, have received a bachelor's degree, have a C average or better if an undergraduate, and be physically qualified.

Leadership is emphasized and courses include Marine Corps History, Customs and Courtesy, Military Law, Drill and Ceremonies, Inspections, and Personal Development. Upon successful completion of training at Quantico, college graduation, and reaching age twenty-one, you can be commissioned a second lieutenant in the Marine Corps and enter a nine-week basic course at Quantico before reporting for duty at a specialist school. Your military obligation upon commissioning is two and one-half years. Women Marine officers are stationed throughout the U.S. and overseas. Assignments include weather forecasting, control tower operations, food service, personnel, administration, public information, teaching, intelligence, supply, and computer operations.

Courses are usually not more than six months in length; however, the language courses vary from two to fifteen months and the electronic officer's (maintenance) course is one year.

The Defense Language Institute trains military personnel in foreign languages; some fifty languages are available, ranging from nineteen to sixty weeks in length. Communications courses cover duties of a communications watch officer, communications procedures, and cryptography. Also offered are electronics courses.

Opportunities for graduate work continue throughout an officer's career. Senior service colleges attended by Navy and Marine Corps officers bring them up to date on international affairs: the Naval War College, Newport, Rhode Island; the Armed Forces Staff College, Norfolk, Virginia; the Industrial College of the Armed Forces, Fort McNair, Washington, D.C.; and the National War College, also at Fort McNair.

Enlistees must complete a Navy Training Course before becoming eligible for promotion to petty officer, and must show ability to perform the duties of the next higher pay grade to become eligible to compete for advancement. Among educational programs is the Secretary of the Navy Scholarship Program for Marines. Advanced degree programs are also available. An enlisted person applying for a scholarship to obtain an undergraduate degree must not have reached age twenty-seven before completing all requirements for the degree.

The Broadened Opportunity for Officer Selection and Training, known as BOOST, is a one-year preparation to enable the Marine to be competitive for selection to the Marine Corps Enlisted Commissioning Education Programs (MECEP), the U.S. Naval Academy, or the Naval Reserve Officer Training Corps (NROTC).

To be eligible, an active-duty enlisted Marine must be a lance corporal or above with one year of active service after completion of recruit training by July 1 of the year in which application is made. The applicant must be at least eighteen but not have reached twenty-four as of July 1 of the year in which assignment to the program is made; must be a high school graduate or have passed the GED high school level tests; must possess a minimum prescribed on the General Classification Test, and must be recommended by the commanding officer.

Marines in the BOOST program receive pay and allowances, promotional opportunities, and normal leave but must agree to obligated service in the Regular Marine Corps.

The *College Degree Program* enables selected Marine officers to earn a bachelor's degree by attending a college or university as a full-time student. It is limited to Regular officers or Reserve officers with career status in the grades of warrant officer to lieutenant colonel who already have enough college credits to earn a bachelor's degree within a maximum period of eighteen months. They receive regular pay and allowances but must pay for tuition, books, and similar expenses. Such officers also have a four- or three-year service obligation after receiving the degree.

The *Navy-Marine four-year college program* is a nonscholarship program for college students who wish to serve as Reserve officers in the Navy or Marine Corps for a specified period. Students are selected from freshmen already enrolled at participating NROTC colleges or universities.

Almost no restriction is placed on undergraduate academic courses chosen, provided they lead to the bachelor's degree. All that

DEFENSE DEPT. PHOTO

A lance corporal hooks up the electrical wiring on an engine under testing at the Marine Corps Supply Center, Barstow, California.

is required prior to graduation is the successful completion of Naval Science and certain university courses and one short summer training session at sea. In return the student receives uniforms and Naval Science textbooks and, on attaining advance NROTC status in the junior and senior years, a tax-free allowance of $100 a month for a maximum of twenty months. Upon graduation and completion of Naval NROTC requirements, the student is commissioned an ensign in the Naval Reserve or a 2nd Lieutenant in the Marine Corps Reserve. The service obligation is three years.

The *Defense Activity for Non-Traditional Education Support*

(DANTES) program provides support to the voluntary education programs of all the military services.

The Assistant Secretary of Defense (Manpower, Installations, and Logistics) exercises policy control over DANTES; the Navy is assigned Executive Agent responsibilities.

The philosophy of DANTES is partnership between civilian and military education in order to make educational opportunities of the highest quality available to men and women of the Army, Navy, Marine Corps, Air Force, Coast Guard, and their National Guard and Reserve components.

Non-traditional education simply means that the educational experience did not take place in the formal classroom, and such experience can become practical components in the educational goals of servicemen and women through examination programs. Thus high school equivalency, college admission, and professional certification programs are available to service personnel throughout the world.

Naval Officers' Graduate Education Program

Technical Curricula

Curriculum	*Location*	*Degree Attainable*
Advanced Science	Naval Postgraduate School (NPGS), Monterey (1 year), followed by study at selected civilian institutions	M.S./Ph.D.
Aerospace Engineering	NPGS, Monterey	M.S./Ae.E.
Aerospace Engineering	California Institute of Technology, M.I.T., Princeton University, Cranfield Institute of Technology (U.K.), Stanford University, North Carolina State University, University of Illinois, Columbia University, Georgia Institute of Technology	M.S.
Civil Engineering (Advanced)	Illinois, Michigan, M.I.T., Stanford, Princeton, etc.	M.S.
Communications Engineering	NPGS, Monterey	B.S./M.S.
Electrical Engineering	University of Michigan	M.S.
Engineering Electronics	NPGS, Monterey	B.S./M.S.
Engineering Electronics	University of Michigan	M.S.
Engineering Science	NPGS, Monterey	B.S.
Environmental Science (Meteorology and Oceanography options)	NPGS, Monterey	B.S./M.S.

Hydrographic Engineering (Geodesy)	Ohio State University	M.S.
Management/Data Processing	NPGS, Monterey	M.S.
Management and Industrial Engineering	Rensselaer Polytechnic	M.S.
Mechanical Engineering	Rensselaer Polytechnic	M.S.
Naval Construction and Engineering	M.I.T. and Webb Institute of Naval Architecture	M.S./Nav.E.
Naval Mechanical and Electrical Engineering	NPGS, Monterey	B.S./M.S.
Nuclear Engineering	M.I.T. (Many include 6-month course at Westinghouse Electric)	M.S.
Nuclear Engineering (Effects)	NPGS, Monterey	B.S./M.S.
Nuclear Power Engineering	University of Michigan, Penn State University	M.S.
Oceanography	NPGS, Monterey; University of Miami (Florida); Texas A&M; and University of Washington	B.S./M.S.
Operations Analysis	NPGS, Monterey	B.S./M.S.
Petroleum Engineering	University of Kansas	M.S.
Weapons Systems	NPGS, Monterey; Michigan State University	B.S./M.S.
Ph.D. Studies	(Location and length of study determined subsequent to selection of candidates)	Ph.D.
Nontechnical Curricula		
Business Administration	Harvard and Stanford Universities	M.B.A.
Defense Intelligence	Defense Intelligence School	None
Financial Management	George Washington University	M.B.A.
International Relations	American and Harvard Universities	M.A.
Logistics (APIT Course)	APIT, Wright-Patterson APB, Ohio	M.S.
Naval Management	NPGS, Monterey	M.S.
Petroleum Administration and Management	Southern Methodist University	M.S.
Petroleum Management	University of Kansas	M.S.
Political Science	Fletcher School of Law and Diplomacy, Tufts University	M.A.
Procurement Management	University of Michigan	M.B.A.
Public Relations	University of Michigan	M.S.
Retailing	Michigan State University	M.B.A.
Subsistence Technology	Michigan State University	M.B.A.
Systems Inventory Management	Harvard University	M.B.A.
Transportation Management	Michigan State University	M.B.A.
Ph.D. Studies	(Location and length of study determined subsequent to selection of candidates)	Ph.D.

Marine Corps Occupations

Career fields are listed alphabetically, beginning with those combat-related groups closed to women (indicated by "C").

Career fields	Duties & Responsibilities	Qualifications	Examples of Civilian Jobs
Field Artillery (C)	Maintains heavy mortars, 175mm, 155mm, 8-inch and 105mm guns.	Mathematical reasoning, mechanical aptitude, good vision and stamina; mechanics, electricity, meteorology and mathematics.	Surveyor, geodetic computer, meteorologist, radio operator, recording engineer and ordnance inspector.
Infantry (C)	Performs as rifleman, machine gunner, or grenadier; infantry unit leader, supervises training and operations of infantry units.	Verbal and mathematical reasoning, good vision and stamina; general mathematics, mechanical drafting, geography and mechanical drawing.	Firearms assembler, gunsmith, policeman, immigration inspector and plant security policeman.
Tank and Amphibian Tractor (C)	Performs as driver, gunner and loader in tanks, armored amphibious tractors. drawing.	Mechanical ability and stamina; auto mechanics, machine shop electricity and mechanical drawing.	Automotive mechanic, bulldozer operator or repairman, caterpillar repairman, armament machinist-mechanic and gunsmith assistant.
Aircraft Maintenance	Performs the mechanical functions of maintenance, repair and modification of Marine air and ground support equipment.	Mechanical or electrical aptitude with manual dexterity, shop mathematics desirable.	Aircraft mechanic, electrician or hydraulics specialist, aviation machinist, sheet metal worker, aircraft instrument maker, repairer.

Career Fields	Duties & Responsibilities	Qualifications	Examples of Civilian Jobs
Airfield Services	Maintains aircraft log books, publications and flight operations records; prepares reports and schedules, installs and repairs aircraft launching and recovery equipment.	Typing, geography and mechanical drawing useful.	Airplane dispatch clerk, flight dispatcher, timekeeper and airport crash truck driver.
Air Traffic Control and Enlisted Flight Crews/Air Support/Anti-Air Warfare	Operates airfield control tower and radio-radar air traffic control systems; serves as navigator, radio and radar operator, intercept controller and anti-air warfare missile batteryman.	Clear speaking voice, good hearing and better than average eyesight; speech, mathematics and electricity; and experience as a ham radio operator helpful.	Airport control tower, or flight radio operator, navigator, instrument-landing truck operator, radio or television studio engineer.
Ammunition and Explosive Ordnance Disposal	Inspects, issues and supervises storage of ammunition and explosives; locates, disarms, detonates or salvages unexploded bombs.	Mechanics, general science, and chemistry useful.	Firearms and ammunition proof director, ordnance technician (government), powder and explosives inspector.
Auditing, Finance and Accounting	Prepares and audits personnel pay records, processes public vouchers and administers and audits unit fiscal accounts.	Computational work, mathematical and attention to detail; typing, bookkeeping, office machines and mathematics useful.	Payroll or cost clerk, bookkeeper, cashier, bank teller, accounting and audit clerk and accountant.
Aviation Ordnance	Maintains and repairs aircraft armament systems, gun pods, machine guns, bomb racks and rocket/missile launcher equipment; assembles and loads bombs, rockets, guns and missiles; handles and stores aviation type munitions.	Electricity, hydraulics and mechanics shop courses useful.	Firearms assembler, gunsmith, armament mechanic and aircraft accessories repairer.

Avionics	Installs and repairs aircraft electrical, communications/navigation and fire control equipment and systems, and air launched guided missiles; serves as electrician and instrument repairman. Repairs and calibrates related test equipment.	Mathematics and shop courses in electricity, hydraulics and electronics helpful.	Radio and television or electrical instrument repairer, communications, electrical or electronics engineer and radio operator.
Band	Performs in Marine Corps Band, unit bands and drum and bugle corps; repairs musical instruments.	Music experience as a member of a high school band or orchestra.	Musician, music librarian, music teacher, bandmaster, orchestra or music director and musical instrument repairer.
Engineer/ Construction Equipment and Shore Party	Performs metal-working, operation and maintenance of fuel storage area, heavy engineering and pioneer equipment, construction and repair of military facilities.	Automotive mechanics, sheet metal working machine shop, carpentry and mechanical drafting useful.	Sheet metal worker, engineering equipment mechanic, carpenter, road machinery operator, rigger and construction superintendent.
Data/ Communications Maintenance	Installs, tests and repairs telephone, teletype and crypto-graphic equipment and cables, calibrates precision electronic, mechanical, dimensional and optical test instruments.	Mathematics, electricity and blueprint reading courses helpful.	Telephone installer and trouble shooter, radio repairer, cable splicer, and office machine serviceperson.
Data Systems	Operates and programs data processing equipment.	Clerical aptitude, manual dexterity and eye-hand coordination, mathematics, accounting and English useful.	Computer operator, programmer and data control coordinator.

Career Fields	Duties & Responsibilities	Qualifications	Examples of Civilian Jobs
Drafting, Surveying, and Mapping	Makes architectural and mechanical drawings, prepares military maps, makes topographic maps, creates or copies articles or illustrative materials.	Mathematics, mechanical drawing and drafting, geography and commercial art helpful.	Architectural or mechanical draftsperson, surveyor or cartographer, geodetic computer, illustrator and commercial artist.
Electronic Maintenance	Installs, tests and repairs air-search radar, radio, radio relay, missile fire control and guidance systems.	Electronics, mathematics, electricity, and blueprint reading useful.	Radio and television repair, radio engineer, electrical instrument repairer, recording communications, and electrical engineer.
Food Service	Performs as cook, baker or meat cutter.	Hygiene, biology, chemistry, home economics and bookkeeping courses useful.	Cook, chef, baker, meat cutter or butcher, caterer, executive chef, dietician and restaurant manager.
Intelligence	Collects, records, evaluates and interprets information,makes detailed study of aerial photographs, conducts interrogations in foreign languages, translates written material and interprets conversations.	Geography, history, government, economics, English, foreign languages, typing, mechanical drafting and mathematics beneficial.	Investigator, research worker, intelligence analyst (government), map maker, cartographic aide and records analyst.
Legal Services	Prepares legal documents, operates stenotype machines.	Manual dexterity; English, filing, typing and shorthand desirable.	Law clerk, court reporter, chief clerk and stenotype operator.
Logistics	Performs administrative duties involving the supply, quartering, movement and transport of Marine units by land, sea and air.	Clerical aptitude, knowledge of verbal and math reasoning; operate office machines, read maps.	Inventory or shipping clerk, pier superintendent, stock control clerk or supervisor and warehouse manager.

Marine Corps Exchange	Keeps and audits books and financial records, performs sales and merchandise stock control duties.	Typing, bookkeeping, business arithmetic, office machines and accounting useful.	Salesman, stock control supervisor, buyer, bookkeeper, accounting clerk, accountant and auditor.
Military Police and Corrections	Enforces military orders, guards military and war prisoners and controls traffic.	Sociology and athletics helpful.	Policeman, ballistics expert and investigator.
Motor Transport	Performs auto mechanics and body repair, motor vehicle and amphibian truck operation.	Automotive mechanics, machine shop, electricity and blueprint reading useful.	Mechanic or automobile body, electrical system repairer, truck driver, motor vehicle dispatcher, and motor transport depot master.
Nuclear, Biological and Chemical	Performs routine duties incident to applying detection, emergency and decontamination measures to gassed or radioactive areas. Inspects and performs preventive maintenance on chemical warfare protection equipment.	Must not have any known hypersensitivity to the wearing of protection clothing; be emotionally stable; biology and chemistry background beneficial.	Laboratory assistant (nuclear, biological or chemical), exterminator, decontaminator.
Operational Communications	Lays communication wire; installs and operates radio, radio telegraph and radio relay equipment; encodes and decodes messages.	Mathematics, typing, electricity and electronics useful.	Radio operator, telephone lineperson, radio broadcasting traffic manager and communications engineer.
Ordnance	Inspects, maintains and repairs infantry, artillery, and anti-aircraft weapons; fire control optical instruments; operates machine tools or modifies metal parts.	Mathematical, mechanics, machine shop and blueprint reading, welding and heat treatment of metal and electricity.	Armament mechanic, gunsmith, time-recording equipment serviceperson, electrician, optical instrument inspector and electrical engineer.

Career Fields	Duties & Responsibilities	Qualifications	Examples of Civilian Jobs
Personnel and Administration	Performs as personnel classifiction clerk, administrative specialist and postal clerk.	Reasoning and verbal ability, and clerical aptitude. English composition, typing, shorthand and social studies helpful.	Secretary-typist, vocational advisor, employment interviewer, manager, office manager, job analyst and postal clerk.
Printing and Reproduction	Performs letterpress and lithographic offset printing; sets type; operates linotype machines, presses, process cameras and bookbinding equipment.	General mathematics, printing and other graphic arts useful.	Printing compositor, linotype operator, photolithographer, press operator, printing bookbinder, printing plant makeup worker and proofreader.
Public Affairs	Gathers material for, writes and edits news stories and historical reports, gathers material for, prepares and edits radio and television broadcast scripts.	English grammar and composition, typing, speech and journalism courses helpful.	News reporter-correspondent, columnist, copyreader, copy or news editor, radio-television announcer and script writer.
Signals Intelligence/ Ground Electronic Warfare	Performs routine duties incident to collecting, translating, recording and disseminating information associated with military plans and operations.	English composition, geography and mathematics beneficial.	Radio intelligence operator, intelligence analyst, investigator and records analyst.
Supply Administration and Operations	Administration procurement, subsistence, packaging and warehousing; requisitions, purchases, receipts, accounts, classifies, stores, issues, sells, packages, preserves and inspects new, scrap, salvage, waste material, supplies and equipment.	Typing, bookkeeping, office machine operation and commercial subjects helpful.	Shipping, receiving, stock and inventory clerk, stock control supervisor, warehouse manager, parts and purchasing agent.

Training and Audiovisual Support	Operates still, motion picture, and aerial cameras; develops film and prints photographs; repairs cameras and edits motion picture films; performs as illustrator or draftsperson.	Mathematics, chemistry and shop courses in electricity; normal color perception desirable.	Commercial illustrator, photographer, cinematographer, copy cameraperson, motion picture film editor, camera and instrument repairer.
Transportation	Handles cargo and transacts business of freight shipping and receiving and passenger transportation.	Typing, bookkeeping, business arithmetic, office machine operation, and commercial subjects beneficial.	Shipping clerk, cargo handler, traffic rate clerk, freight traffic, passenger and railroad station agent.
Utilities	Installs, operates, and maintains electrical, water supply, heating, plumbing, sewage, refrigeration, hygiene and air conditioning equipment.	Mechanical aptitude and manual dexterity important; vocational school shop course in industrial arts and crafts beneficial.	Electrician, plumber, steam fitter, refrigeration mechanic, electric motor repairer and stationary engineer.
Weather Service	Collects, records and analyzes meteorological data; makes visual and instrumental observations and enters them on appropriate charts; forecasts short, intermediate and long range weather conditions.	Visual acuity correctable to 20/20, normal color perception; mathematics desirable, meteorology and astronomy helpful.	Meteorologist, weather forecaster and observer.

Chapter IV

The United States Air Force

"With dauntless resolution and unconquerable faith"

—from the monument to the Wright Brothers' first flight in Kitty Hawk, North Carolina

On April 1, 1954, President Eisenhower signed the bill authorizing establishment of the United States Air Force Academy. The site chosen is 10 miles north of Colorado Springs, Colorado, on the slopes of the Rampart Range of the Rocky Mountains. Pikes Peak rises 14,000 feet behind the academy site.

The Air Force Academy covers 18,000 acres of former ranch land divided into five mesas with valleys in between. The Academy buildings are designed in contemporary architectural style featuring glass, aluminum, steel, and white marble. Dramatic and simple, the buildings blend into the spectacular landscape. The Cadet Chapel with seventeen towering spires is a world-famous multidenominational house of worship. South of the cadet complex are two housing developments for officers and airmen, a shopping center, a hospital, the Academy Preparatory School, a supply and service center, and an airstrip for cadet flying activities. A football stadium, a golf course, and recreation areas in the mountains were built with private funds donated through the Air Force Academy Foundation.

The first class was graduated on June 3, 1959, and at last count there were some 4,000 members of the Cadet wing. Like West Point and Annapolis, this newest of the military academies is developing its own traditions. Its Eagle statue has become a landmark. The motto over the Academy archway reads "Bring Me Men." The motto remains even though women cadets were authorized admission to the Academy in 1976.

The four-year Academy program leads to a Bachelor of Science degree and a commission in the Regular Air Force. Instruction begins with basic cadet training, a rigorous orientation to military life, during the new cadet's first summer. Other summer programs give cadets an insight into the operational Air Force as well as other branches of the armed forces. Through optional programs, cadets have opportunities to learn skills in light plane flying, parachuting, navigation, and soaring. Cadets qualified to fly can volunteer for flight training as they begin the First Class year. They learn to fly T-41 aircraft, and after graduation they can earn their wings in the Air Force. Navitagor-qualified cadets can also select instruction to prepare for postgraduate training.

Academic Program

The Eagle statue, donated by the Air Training Command, commemorates the opening of its permanent cadet facilities in 1959. The inscription thereon, "Man's Flight Through Life Is Sustained by the Power of His Knowledge," symbolizes the academic program as it implements the Air Force's flying mission, which is to develop well-rounded, broadily educated professional officers. The Academy program combines academic knowledge, leadership and airman-

U.S. AIR FORCE PHOTO

A member of a fighter wing attaches an electrical cord to an outlet on a Thunderbolt II aircraft to perform maintenance.

ship, physical education, and athletics. It provides each cadet with an education in primary areas of knowledge and also in the specialization needed to perform effectively as an Air Force officer. Basic courses in the sciences, social sciences, and humanities are required during the first two years. Advanced courses oriented toward Air Force career fields are taken during the last two years.

The Air Force needs specialists in many areas, so individual aptitudes can be taken into account. A broad range of major areas of study (plus elective courses and enrichment programs) enables a cadet to advance beyond the prescribed academic program. For those who apply themselves and have the aptitude and ability, opportunities to attend graduate school come early.

Physical Education and Athletics

Men and women cadets undergo the same type of required physical and military training. During the summer program, the cadets develop strength, endurance, agility, and coordination as well as a sense of teamwork and competition. Strenuous conditioning exercises, endurance tests, and obstacle courses are on the program, as well as competitive sports, Field Day events, and field encampment training. It is a rugged test of stamina, and new cadets find the program physically and mentally taxing. But it is all for a purpose: to build a high level of performance and ability to handle the demands of cadet life.

Physical education is a part of all four years at the Academy and includes fitness methods, aerobic conditioning, skills development, and sports participation. Cadets improve swimming skills and learn lifesaving and survival techniques, self-defense in unarmed combat courses, and sports that will be enjoyable in future years such as golf, tennis, handball, volleyball, and squash. An intercollegiate sports schedule includes the service academies and leading universities; cadets not participating in varsity events are required to participate in intramurals on campus.

Cadet Activities

After the first year, cadets have more leisure time for extracurricular activities. Over seventy organized activities are available, including a ski club providing instruction and trips to areas in the Rocky Mountain region, and a balloon club. Both men and women cadets participate in extracurricular activities on a coed basis.

The Academy also provides recreational facilities for cadets when they have weekend privileges. The Lawrence Paul area, on a small lake within easy walking distance of the cadet area, is used for fishing, dances, parties, and picnics. The Farish Memorial recreational area on a lake in the mountains four miles west of the Academy has accommodations for fishing, horseback riding, ice skating, boating, and barbecues. Arnold Hall is the cadet social center for class dances and formal balls.

Entrance Requirements

To be eligible for the Air Force Academy, you should plan ahead while you're in high school. In the academic area, complete four credits in English and four in mathematics, with the remainder of your credits in the sciences, social sciences, and foreign language. Your grade record will be important in qualifying you for a cadet appointment.

Take part in extracurricular activities to show your leadership potential. Athletics, public speaking, class office, Scouts, Air Force Reserve Officer Training Corps, and Civil Air Patrol participation all help to develop leadership qualities.

Physical fitness counts. Take care of your health and condition yourself through vigorous team sports.

A candidate for the Air Force Academy must be at least seventeen years of age and not have reached twenty-two on July 1 of the year of entry; be unmarried; be of good moral character; be a United States citizen, unless applying as an Allied Student from one of the other American republics; and be in good physical shape.

In the second semester of your junior year in high school, you are eligible to request a Precandidate Questionnaire. Write a letter, giving your age and the year you expect to graduate from high school, to:

Admissions Liaison Office
USAF Academy, CO 80840

Also in your junior year, you should request a nomination from both of your U.S. Senators and the Representative for your congressional district. You need not know them personally to apply. Your school guidance counselor can give you the name of your Air Force Academy Liaison Officer.

Air Force for Women

Of the numerous Air Force career fields, the only ones excluding women because of the combat prohibition are Defensive Aerial Gunnery Helper; Flight Engineer Helper—Helicopter-qualified; Pararescue Recovery Helper, Combat Control Helper, and Tactical Air Command and Control Helper.

All other career fields are open to women officers: chaplain, legal, medical, line officers—some of whom are pilots and navigators. By

U.S. AIR FORCE PHOTO

A technical sergeant of the 32nd Aeromedical Evacuation Group arrives at Travis Air Force Base, California, during an aeromedical evacuation training exercise.

U.S. AIR FORCE PHOTO

A group of officers arrive at Mather Air Force Base, California, to begin training in the school of navigation, previously an all-male school.

the end of 1987, the number of women in the Air Force is projected to reach 70,000.

If you are uncertain about your aptitudes, the Air Force will provide a Vocational Aptitude Battery exam free and without obligation. The test might be useful to you in planning your career, as it measures aptitudes in the following fields; mechanical, administrative, general occupations, and electronics.

Air Force Reserve

You're finishing high school and are facing a time of decision-making. An option many young people overlook is one of the best part-time jobs in America—a position with the Air Force Reserve. You serve your country and your community for two days a month plus two weeks of active duty training each year in return for these rewards:

Salary. After you've completed your school or on-the-job training you earn a very competitive salary. Pay is adjusted periodically for increase in the cost of living. The current system also provides salary increases for each two years of service and additional pay if you are on flying status.

Personal recognition. Your Air Force accomplishments are recognized through awards, decorations, and promotions.

Camaraderie. You'll make new friends and associates in the Air Force Reserve.

Leadership experience. Training and experience that can benefit you in your regular job.

Insurance. You can purchase a substantial amount of insurance for a small monthly fee, and you're automatically covered for hospitalization and medical care while on active duty.

Retirement benefits. A noncontributory retirement program that may be worth several hundred dollars a month to you on retirement.

Opportunities for travel—with your Air Force Reserve Unit or if assigned to a flying organization during your unit training weekend.

Uniforms are supplied once you're assigned to a unit.

Reemployment rights. Reservists are entitled to return to their regular jobs if ordered to active duty during a national emergency.

More than 80 percent of Air Force Reserve skills have civilian counterparts—skills such as jet engine mechanics, computer programming and repair, telephone repair, plumbing, and carpentry. Among specialties are Air Cargo Specialist, Aircraft Mechanic, Avionics/Electronics, Fire Fighter, Loadmaster, and Medical Service Technician.

U.S. AIR FORCE PHOTO

A Command and Control specialist updates the fighter aircraft status board during a cold-weather training exercise at Elmendorf Air Force Base, Alaska.

To be eligible to enlist in the Air Force Reserve, you must be at least seventeen and not more than thirty-five years of age (under eighteen requires parental consent); pass the Armed Services Vocational Aptitude Battery (ASVAB); be of good moral character; and be in good physical condition. Currently, Air Force Reserve enlistments are for six years.

The Air Force Reserve also accepts doctors and nurses, for whom there are special requirements.

Air Force Junior ROTC

To be eligible to enroll in the Air Force ROTC, you must be at least fourteen years of age, a citizen of the United States, and physically fit. You may normally enter AFJROTC in tenth grade. Uniform, insignia, and textbooks are furnished.

The first year introduces students to modern aircraft, spacecraft, the history of flight, and the emergence of aerospace careers.

The second year teaches principles of flight, propulsion, and navigation. The final year is a more in-depth study of space operations and exploration, international cooperation in space, the human body in flight, and the role of the military in U.S. history.

All this is designed to develop leadership qualities and management and speaking skills. You also learn military customs and courtesies, flag etiquette, and basic drill positions and commands.

It is not all work. You can attend dances and military balls sponsored by your Junior ROTC unit, be part of a color guard or drill team at athletic events, and visit Air Force bases, airports, and aerospace industries.

Students who complete three years of Air Force Junior ROTC and enroll in Air Force ROTC in college may receive credit for the first year of the four-year college program. The hosting high school can nominate qualified candidates from its AFJROTC for an Air Force Academy appointment. If you enlist in the military, successful completion of the three-year Aerospace Education Program will enable you to enter at two pay grades higher than other enlistees.

COLLEGES and UNIVERSITIES AFFILIATED with AIR FORCE ROTC

Air Force ROTC detachment locations are indicated below by the three-digit numbers before the school names; some have three digits followed by one letter. Students at schools preceded by a "+" (listed below detachments) attend Air Force ROTC classes at the detachment. For example, Detachment 012 is Samford University; students at Birmingham Southern College attend Air Force ROTC classes at Samford.

ALABAMA

005 Auburn University, Auburn 36849 (205) 826-4355
010 University of Alabama, University 35486 (205) 348-5900
012 Samford University, Birmingham 35209 (205) 870-2859
+ Birmingham Southern College, Birmingham 35204
+ Miles College, Birmingham 35208
+ University of Alabama in Birmingham, Birmingham 35294
+ University of Montevallo, Montevallo 35115
015 Tuskegee Institute, Tuskegee 36088 (205) 727-8384
017 Troy State University, Troy 36082 (205) 566-5115
019 Alabama State University, Montgomery 36195 (205) 293-4306
019 Alabama State University, Montgomery 36195 (205) 293-4306
+ Auburn University at Montgomery, Montgomery 36117
+ Huntingdon College, Montgomery 36106

ARIZONA

20 University of Arizona, Tucson 85721 (602) 621-3521
025 Arizona State University, Tempe 85281 (602) 965-3181
+ Grand Canyon College, Phoenix 85017
027 Northern Arizona University, Flagstaff 86011 (602) 523-5371
028 Embry-Riddle Aeronautical at Prescott, Prescott 86301 (602) 778-4130

ARKANSAS

030 University of Arkansas, Fayetteville 72701 (501) 575-3651

CALIFORNIA

035 California State University, Fresno 93740 (209) 291-9947
+ West Coast Bible College, Fresno 93710
045 San Jose State University, San Jose 95192 (408) 277-2743
+ Ohlone College, Fremont 94538
+ San Jose City College, San Jose 95128
+ Stanford University, San Jose 94305
+ University of Santa Clara, Santa Clara 95053
055 University of California at Los Angeles, Los Angeles 90024 (213) 825-1742
+ California Lutheran College, Thousand Oaks 91360
+ California State College, San Bernardino 92407
+ California State University, Dominguez Hills, Carson 90747
+ California State Polytech University of Pomona, Pomona 91768
+ California State University at Fullerton, Fullerton 92634
+ California State University at Long Beach, Long Beach 90840
+ California State University at Los Angeles, Los Angeles 90032
+ California State University at Northridge, Northridge 91330
+ Los Angeles Mission College, San Fernando 91340
+ Mount St Mary's College, Los Angeles 90049
+ Northrop University, Inglewood 90306
+ Santa Monica College, Santa Monica 90405
+ Southern Illinois University, Sunnymead 92388
+ University of California Irvine, Irvine 92717
+ University of California Riverside, Riverside 92521
+ University of LaVerne, LaVerne 91750
+ University of California Santa Barbara, Santa Barbara 93106
055A Loyola Marymount University, Los Angeles 90045 (213) 670-0113
+ California State University Dominguez Hills, Carson 90747
+ California State University, Long Beach, Long Beach 90840
+ California State College, San Bernardino 92407
+ California State Polytech University at Pomona 91768
+ California State University at Los Angeles, Los Angeles 90032
+ California State University at Fullerton, Fullerton 92634
+ California State University Northridge, Northridge 91330
+ East Los Angeles College, Monterey Park 91754
+ Los Angeles Harbor College, Wilmington 90744
+ Los Angeles Pierce College, Woodland Hills 91371
+ Los Angeles Trade Tech College, Los Angeles 90015
+ Los Angeles Valley College, Van Nuys 91401
+ Mount St Mary's College, Los Angeles 90049
+ Northrop University, Inglewood 90306
+ Pepperdine University, Malibu 90265

+ University of California Irvine, Irvine 92717
+ University of Redlands, Redlands 92373
+ University of California Riverside, Riverside 92521
+ Westmont College, Santa Barbara 93108

055B California State University at Long Beach, Long Beach 90503 (213) 498-5743
+ California State College, San Bernardino 92407
+ California State University, Dominguez Hills, Carson 90747
+ California State University at Fullerton, Fullerton 92634
+ California State University Los Angeles, Los Angeles 90032
+ California State Polytech University at Pomona, Pomona 91768
+ University of California Riverside, Riverside 92521

060 University of Southern California, Los Angeles 90089 (213) 743-2670
+ Biola College, La Mirada 90639
+ California Institute of Technology, Pasadena 91125
+ California Lutheran College, Thousand Oaks 91360
+ California State College, San Bernardino 92407
+ California State Polytech University at Pomona, Pomona 91768
+ California State University, Dominguez Hills, Carson 90747
+ California State University at Fullerton, Fullerton 92634
+ California State University at Los Angeles, Los Angeles 90032
+ California State University at Northridge, Northridge 91324
+ Chapman College, Orange 92666
+ Claremont Men's College, Claremont 91711
+ Harvey Mudd College, Claremont 91711
+ Northrop University, Inglewood 90306
+ Occidental College, Los Angeles 90041
+ Pepperdine University, Malibu 90265
+ Pomona College, Claremont 91711
+ University of California, Irvine 92717
+ University of California Riverside, Riverside 92521
+ Whittier College, Whittier 90608

075 San Diego State University, San Diego 92182 (619) 265-5545
+ Point Loma College, San Diego 92106
+ University of California San Diego, La Jolla 92093
+ University of San Diego, San Diego 92110
+ National University, San Diego 92108

085 University of California at Berkeley, Berkeley 94720 (415) 642-3572
+ California State University at Hayward, Hayward 94542
+ City College of San Francisco, San Francisco 94112
+ Holy Names College, Oakland 94619
+ Mills College, Oakland 94613
+ Ohlone College, Fremont 94538
+ Sonoma State College, Rohnert Park 94928
+ St Mary's College, Moraga 94575

088 California State University at Sacramento, Sacramento 95819 (916) 454-7315
+ American River College, Sacramento 95841
+ Consumnes River College, Sacramento 95823
+ Sacramento City College, Sacramento 95822

+ Sierra College, Rocklin 95677
+ University of California at Davis, Davis 95616
+ University of Pacific, Stockton 95211

COLORADO

090 Colorado State University, Fort Collins 80523 (303) 491-6476
100 University of Northern Colorado, Greeley 80639 (303) 351-2061
105 University of Colorado, Boulder 80309 (303) 492-8351
+ Colorado School of Mines, Golden 80401
+ Colorado Women's College, Denver 80220
+ Metropolitan State College, Denver 80204
+ Regis College, Denver 80221
+ University of Colorado at Denver, Denver 80202
+ University of Denver, Denver 80208

CONNECTICUT

115 University of Connecticut, Storrs 06268 (203) 486-2225
+ Central Connecticut State, New Britain 06050
+ Eastern Connecticut State, Willimantic 06226
+ Southern Connecticut State, New Haven 06515
+ Trinity College, Hartford 06106
+ University of Hartford, West Hartford 06117
+ Western Connecticut State College, Danbury 06810

DELAWARE

128 University of Delaware, Newark 19716 (302) 738-2863
+ Wilmington College, New Castle 19720

DISTRICT OF COLUMBIA

130 Howard University, Washington 20059 (202) 636-6788
+ American University, Washington 20016
+ Georgetown University, Washington 20057
+ George Washington University, Washington 20052
+ The Catholic University of America, Washington 20064
+ Trinity College, Washington 20017
+ University of the District of Columbia, Mt Vernon Square, Washington 20005

FLORIDA

145 Florida State University, Tallahassee 32306 (904) 644-3461
+ Florida A&M University, Tallahassee 32307
150 University of Florida, Gainesville 32611 (904) 392-1355
155 University of Miami, Coral Gables 33124 (305) 284-2870
+ Barry College, Miami 33161
+ Biscayne College, Miami 33054
+ Florida Memorial College, Miami 33054
157 Embry-Riddle Aeronautical University, Daytona Beach 32014 (904) 253-4089
+ Bethune-Cookman College, Daytona Beach 32015
+ University of Central Florida, Daytona Center 32014
158 University of South Florida, Tampa 33620 (813) 974-3367 ext 261
+ Saint Leo College, Saint Leo 33753
+ University of Tampa, Tampa 33606

159 University of Central Florida, Orlando 32812 (305) 275-2264
+ Florida Southern College, McCoy Campus, Orlando 32809
+ Rollins College, Winter Park 32789

GEORGIA

160 University of Georgia, Athens 30602 (404) 543-4981
165 Georgia Institute of Technology, Atlanta 30332 (404) 984-4175
+ Agnes Scott College, Decatur 30030
+ Clark College, Atlanta 30314
+ Georgia State University, Atlanta 30314
+ Morehouse College, Atlanta 30314
+ Morris Brown College, Atlanta 30314
+ Southern Technical Institute, Marietta 30060
+ Spelman College, Atlanta 30314
172 Valdosta State College, Valdosta 31698 (912) 333-5954

HAWAII

175 University of Hawaii at Manoa, Honolulu 96822 (808) 948-7762
+ Chaminade University of Honolulu, Honolulu 96816
+ Hawaii Pacific College, Honolulu 96814
+ West Oahu College, Pearl City 96782

ILLINOIS

190 University of Illinois, Urbana 61820 (217) 333-1927
195 Illinois Institute of Technology, Chicago 60616 (312) 567-3525
+ Chicago State University, Chicago 60628
+ Elmhurst College, Elmhurst 60126
+ Governors State University, Park Forest South 60466
+ Lewis University, Lockport 60441
+ North Central College, Naperville 60566
+ Northern Illinois University, DeKalb 60115
+ Northeastern Illinois University, Chicago 60625
+ Northwestern University, Evanston 60201
+ Rush University, Chicago 60612
+ Saint Xavier College, Chicago 60655
+ University of Illinois at Chicago Circle, Chicago 60680
205 Southern Illinois University, Carbondale 62901 (618) 453-2481
206 Southern Illinois University at Edwardsville, Edwardsville 62026 (618) 692-3180
+ McKendree College, Lebanon 62254
207 Parks College of St. Louis, Cahokia 62206

INDIANA

215 Indiana University, Bloomington 47401 (812) 335-4191
+ Butler University, Indianapolis 46208
+ DePauw University, Greencastle 46135
+ Indiana University Purdue University at Indianapolis, Indianapolis 46202
+ Marian College, Indianapolis 46202
+ Rose-Hulman Institute of Technology, Terre Haute 47803
218 Indiana State University, Terre Haute 47809
220 Purdue University, West Lafayette 47907 (317) 494-2042
225 University of Notre Dame, Notre Dame 46556 (219) 239-6634

+ Indiana University at South Bend, South Bend 46544
+ St Mary's College, Notre Dame 46556

(295) University of Louisville, Louisville KY 40208 (502) 588-6576
+ Indiana University Southeast, New Albany 47150

IOWA

250 Iowa State University, Ames 50011 (515) 294-1716
+ Drake University, Des Moines 50311

255 University of Iowa, Iowa City 52242 (319) 353-4418

KANSAS

270 Kansas State University, Manhattan 66506 (913) 532-6600

280 The University of Kansas, Lawrence 66045 (903) 864-4676
+ Washburn University, Topeka 66621
+ Mid-America Nazarene College, Olathe 66061

KENTUCKY

290 University of Kentucky, Lexington 40506 (606) 257-1681
+ Eastern Kentucky University, Richmond 40475
+ Georgetown College, Georgetown 40324
+ Kentucky State University, Frankfort 40601
+ Transylvania University, Lexington 40508

295 University of Louisville, Louisville 40292 (502) 588-6576
+ Bellarmine College, Louisville 40205
+ Indiana University Southeast, New Albany IN 47150
+ Spalding College, Louisville 40203

(665) University of Cincinnati, Cincinnati OH 45221 (513) 475-2237
+ Northern Kentucky University, Highland Heights 41076
+ Thomas More College, Fort Mitchell 41017

LOUISIANA

305 Louisiana Tech University, Ruston 71272 (318) 257-2740

310 Louisiana State University and A&M College, Baton Rouge 70893 (504) 388-4407
+ Southern University A&M College, Baton Rouge 70813

311 Grambling State University, Grambling 71245 (318) 247-6768

315 University of Southwestern Louisiana, Lafayette 70504 (318) 235-6361

320 University of New Orleans, New Orleans 70122 (504) 286-6647
+ Dillard University, New Orleans 70122
+ Loyola University in New Orleans, New Orleans 70118
+ Our Lady of Holy Cross College, New Orleans 70114
+ Southern University in New Orleans, New Orleans 70126
+ Tulane University, New Orleans 70118
+ Xavier University of Louisiana, New Orleans 70125

MAINE

326 University of Maine at Orono 04464 (207) 581-1384
+ Husson College, Bangor 04401

MARYLAND

330 University of Maryland, College Park 20742 (301) 454-3245
+ Bowie State College, Bowie 20715
+ George Mason University, Fairfax VA 22030
+ Johns Hopkins University, Baltimore 21218

+ Loyola College, Baltimore 21210
+ Shepherd College, Shepherdstown WV 25443
+ Towson State University, Baltimore 21204
+ Western Maryland College, Westminster 21157

MASSACHUSETTS

340 College of the Holy Cross, Worcester 01610 (617) 793-3343
+ Anna Maria College, Paxton 01612
+ Assumption College, Worcester 01609
+ Central New England College, Worcester 01608
+ Clark University, Worcester 01610
+ Worcester Polytechnic Institute, Worcester 01609
+ Worcester State College, Worcester 01602

345 University of Lowell, Lowell 01854 (617) 459-9301
+ Bentley College, Waltham 02154
+ Daniel Webster College, Nashua, NH 03060
+ Endicott College, Beverly 01915
+ Gordon College, Wenham 01984
+ New England College, Henniker NH 03204
+ New Hampshire College, Manchester NH 03104
+ Notre Dame College, Manchester NH 03104
+ Rivier College, Nashua NH 03060
+ Saint Anselm College, Manchester NH 03102
+ Salem State College, Salem 01970

355 Boston University, Boston 02215 (617) 354-4705
+ Northeastern University, Boston 02115

365 Massachusetts Institute of Technology, Cambridge 02139 (617) 253-4475
+ Harvard University, Cambridge 02138
+ Tufts University, Medford 02155
+ Wellesley College, Wellesley 02181

370 University of Massachusetts, Amherst 01003 (413) 545-2437
+ Amherst College, Amherst 01002
+ Mount Holyoke College, South Hadley 01075
+ Smith College, Northampton 01063
+ Western New England College, Springfield 01119

MICHIGAN

380 Michigan State University, East Lansing 48824 (517) 355-2168

390 The University of Michigan, Ann Arbor 48109 (313) 764-2405
+ Concordia College, Ann Arbor 48105
+ Eastern Michigan University, Ypsilanti 48197
+ Lawrence Institute of Technology, Southfield 48075
+ University of Michigan at Dearborn, Dearborn 48128

400 Michigan Technological University, Houghton 49931 (906) 487-2652

MINNESOTA

410 The College of St Thomas, St Paul 55105 (612) 647-5083
+ Bethel College, Saint Paul 55112
+ Augsburg College, Minneapolis 55454
+ College of St Catherine, St Paul 55105
+ Hamline University, St Paul 55104
+ Macalester College, St Paul 55105

415 University of Minnesota, Minneapolis 55455 (612) 373-2205

420 University of Minnesota at Duluth, Duluth 55812 (218) 726-8159
+ College of St Scholastica, Duluth 55811
+ University of Wisconsin at Superior, Superior 54880
(610) North Dakota State University of A&AS, Fargo ND 58105 (701) 237-8186
+ Concordia College, Moorhead 56560
+ Moorhead State University, Moorhead 56560

MISSISSIPPI

425 Mississippi State University, Mississippi State 39762 (601) 325-3810
+ Mississippi University for Women, Columbus 39701
430 University of Mississippi, University 38677 (601) 232-7357
430A Mississippi Valley State University, Itta Bena 38941 (601) 254-6392
+ Delta State University, Cleveland 38732
432 University of Southern Mississippi, Hattiesburg 39401 (601) 266-4468
+ William Carey College, Hattiesburg 39401

MISSOURI

437 Southeast Missouri State University, Cape Girardeau 63701 (314) 651-2184
440 University of Missouri, Columbia 65211 (314) 882-7621
+ Columbia College, Columbia 65201
+ Stephens College, Columbia 65201
+ William Woods College, Fulton 65251
442 University of Missouri-Rolla, Rolla 65401 (314) 341-4932
(207) Parks College of St Louis University, Cahokia IL 62206
+ Harris Stowe State College, St Louis 63103
+ St Louis University, St Louis 63103
+ University of Missouri at St Louis, St Louis 63121
+ Washington University, St Louis 63130

MONTANA

450 Montana State University, Bozeman 59717 (406) 994-4022

NEBRASKA

465 University of Nebraska, Lincoln 68588 (402) 472-2473
+ Concordia Teachers College, Seward 68434
+ Nebraska Wesleyan University, Lincoln 68504
470 University of Nebraska at Omaha, Omaha 68182 (402) 554-2318
+ Bellevue College, Bellevue 68005
+ College of St Mary, Omaha 68124
+ Creighton University, Omaha 68178
+ University of Nebraska Medical Center, Omaha 68105

NEW HAMPSHIRE

475 University of New Hampshire, Durham 03824
+ Nathaniel Hawthorne College, Antrim 03440
+ New England College, Henniker 03242
+ New Hampshire College, Manchester 03104
+ St Anselm's College, Manchester 03102
+ The University of Southern Maine, Portland ME 04103
+ University of New Hampshire, Plymouth State College, Plymouth 03264
(345) University of Lowell, Lowell MA 01854 (617) 459-9301
+ Daniel Webster College, Nashua 03060
+ New Hampshire College, Manchester 03104

+ Notre Dame College, Manchester 03104
+ Rivier College, Nashua 03060
+ Saint Anselm College, Manchester 03102

NEW JERSEY

485 Rutgers, The State University, New Brunswick 08901 (201) 932-7430
+ Monmouth College, West Long Beach 07764
+ Princeton University, Princeton 08540
+ Rider College, Trenton 08648
+ Rutgers University, Camden Campus 08102
+ Trenton State College, Trenton 08625
+ Wagner College, Staten Island NY 10301

490 New Jersey Institute of Technology, Newark 07102 (201) 645-5240
+ Fairleigh Dickenson University-Teaneck, Teaneck 07666
+ Jersey City State College, Jersey City 07305
+ Kean College of New Jersey, Union 07083
+ Montclair State College, Upper Montclair 07043
+ Rutgers University, Newark Campus 07102
+ Seton Hall University, South Orange 07079
+ Stevens Institute of Technology, Hoboken 07030
+ St Peter's College, Jersey City 07306
+ William Paterson College, Wayne 07470

NEW MEXICO

505 New Mexico State University, Las Cruces 88003 (505) 646-2136

505A University of Texas at El Paso, TX 79958

510 University of New Mexico, Albuquerque 87131 (505) 277-4502
+ University of Albuquerque, Albuquerque 87140

NEW YORK

520 Cornell University, Ithaca 14853 (607) 256-4004
+ Ithaca College, Ithaca 14850
+ SUNY College at Cortland, Cortland 13045
+ Wells College, Aurora 13056

535 Syracuse University, Syracuse 13210 (315) 476-9272
+ LeMoyne College, Syracuse 13214
+ New School for Social Research, New York 10011
+ SUNY College of Environmental Science & Forestry, Syracuse 13210
+ Utica College of Syracuse University, Utica 13502

535B Rochester Institute of Technology, Rochester 14623 (315) 476-9272
+ University of Rochester, Rochester 14627

536 Clarkson College of Technology, Potsdam 13676 (315) 268-7962
+ SUNY College at Potsdam, Potsdam 13767
+ St Lawrence University, Canton 13617

550 Rensselaer Polytechnic Institute, Troy 12181 (518) 266-6236
+ College of St Rose, Albany 12203
+ Russell Sage College, Troy 12180
+ Siena College, Loudonville 12211
+ Skidmore College, Saratoga Springs 12866
+ SUNY at Albany, Albany 12222
+ SUNY Empire State College, Saratoga Springs 12866
+ Union College, Schenectady 12308

560 Manhattan College, Riverdale 10471 (212) 920-0201
+ Adelphi University, Garden City 11530
+ College of Mount St Vincent, Riverdale 10471
+ Columbia University, New York 10027
+ C. W. Post Center of Long Island University, Greenvale 11548
+ Dowling College, Oakdale, Long Island 11769
+ Long Island University Brooklyn Center, Brooklyn 11201
+ Mercy College, Dobbs Ferry 10522
+ Molloy College, Rockville Centre 11570
+ New York Institute of Technology, Old Westbury 11568
+ Pace University, Pace Plaza, New York 10038
+ Polytechnic Institute of New York, Brooklyn 11201
+ Saint Francis College, Brooklyn 11201
+ Saint Joseph's College/Brentwood, Patchogue 11722
+ Saint Thomas Aquinas College, Sparkill 10976
+ SUNY at Old Westbury, Old Westbury 11568
+ SUNY at Stony Brook, Stony Brook 11794
+ Wagner College, Staten Island 10301

NORTH CAROLINA

585 Duke University, Durham 27706 (919) 684-3641
+ North Carolina Central University, Durham 27707
590 University of North Carolina, Chapel Hill 27514 (919) 962-2074
590A University of North Carolina at Charlotte, Charlotte 28223 (704) 597-4537
+ Barber-Scotia College, Belmont 28025
+ Belmont Abbey College, Belmont 28012
+ Davidson College, Davidson 28036
+ Johnson C. Smith University, Charlotte 28216
+ Sacred Heart College, Belmont 28012
+ Queens College, Charlotte 28274
+ Wingate College, Wingate 28174
+ Winthrop College, Rock Hill 29733
595 North Carolina State University at Raleigh, Raleigh 27607 (919) 737-2417
+ Meredith College, Raleigh 27611
+ Shaw University, Raleigh 27611
+ St. Augustine's College, Raleigh 27611
600 East Carolina University, Greenville 27834 (919) 757-6598
605 North Carolina A&T State University, Greensboro 27411 (919) 379-7707
+ Bennett College, Greensboro 27420
+ Greensboro College, Greensboro 27420
+ Guilford College, Greensboro 27410
+ High Point College, High Point 27262
+ University of North Carolina, Greensboro 27412
607 Fayetteville State University, Fayetteville 28301 (919) 486-1465
+ Pembroke State University, Pembroke 28372

NORTH DAKOTA

610 North Dakota State University of A&AS, Fargo 58105 (701) 237-8186
+ Concordia College, Moorhead, MN 56560
+ Moorhead State University, Moorhead, MN 56560

OHIO

620 Bowling Green State University, Bowling Green 43403 (419) 353-6093
+ Ashland College, Ashland 44805
+ Defiance College, Defiance 43512
+ Findlay College, Findlay 45840
+ Heidelberg College, Tiffin 44883
+ Ohio Northern University, Ada 45810
+ University of Toledo, Toledo 43606

630 Kent State University, Kent 44242 (216) 672-2182

640 Miami University, Oxford 45056 (513) 529-2031
+ Miami University Middletown Branch, Middletown 45052

643 Wright State University, Dayton 45435 (513) 529-2031
+ Antioch College, Yellow Springs 45387
+ Central State University, Wilberforce 45384 (513) 873-2730
+ Cedarville College, Cedarville 45314
+ University of Dayton, Dayton 45469
+ Urbana College, Urbana 43078
+ Wilberforce University, Wilberforce 45384
+ Wilmington College, Springfield 45177
+ Wittenburg University, Springfield 45501

645 The Ohio State University, Columbus 43210 (614) 422-5441
+ Capital University, Columbus 43209
+ Franklin University, Columbus 43215
+ Ohio Dominican College, Columbus 43219
+ Ohio Wesleyan University, Delaware 43016
+ Otterbein College, Westerville 43081

650 Ohio University, Athens 45701 (614) 594-6613

660 The University of Akron, Akron 44325 (216) 375-7653

665 University of Cincinnati, Cincinnati 45221 (513) 475-2237
+ Cincinnati Tech College, Cincinnati 45225
+ College of Mount St. Joseph, Mount St. Joseph 45051
+ Edgecliff College, Cincinnati 45207
+ Northern Kentucky University, Highland Heights KY 41076
+ Thomas More College, Fort Mitchell KY 41017
+ Xavier University, Cincinnati 45207

OKLAHOMA

670 Oklahoma State University, Stillwater 74078 (405) 624-4255

675 The University of Oklahoma, Norman 73019 (405) 325-3211
+ Oklahoma City University, Oklahoma City 73106
+ Oklahoma Christian College, Oklahoma City 73111

OREGON

685 Oregon State University, Corvallis 97331 (503) 754-3291
+ Western Oregon State College, Monmouth 97361
+ University of Oregon, Eugene 97403

695 University of Portland, Portland 97203 (503) 283-7216
+ Concordia College, Portland 97211
+ Portland State University, Portland 97207
+ University of Oregon Health Sciences Center, Portland 97201
+ Williamette University, Salem 97301
+ Warner-Pacific College, Portland 97215

PENNSYLVANIA

715 Lehigh University, Bethlehem 18015 (215) 861-3290
+ Allentown College of St Francis Desales, Center Valley 18034
+ Cedar Crest College, Allentown 18104
+ East Stroudsburg State College, East Stroudsburg 18301
+ Kutztown State College, Kutztown 19530
+ Lafayette College, Easton 18042
+ Moravian College, Bethlehem 18018
+ Muhlenburg College, Allentown 18104
+ Pennsylvania State-Allentown, Fogelsville 18051

720 The Pennsylvania State University, University Park 16802 (814) 865-1983

730 University of Pittsburgh, Pittsburgh 15260 (412) 624-6397
+ Carlow College, Pittsburgh 15213
+ Chatham College, Pittsburgh 15232
+ Duquesne University, Pittsburgh 15282
+ LaRoche College, Pittsburgh 15237
+ Point Park College, Pittsburgh 15222
+ Robert Morris College, Corapolis 15108
+ St Vincent College, Latrobe 15650

745 Grove City College, Grove City 16127 (412) 458-8430
+ Slippery Rock State College, Slippery Rock 16057

750 St Joseph's University, Philadelphia 19131 (215) 879-7310
+ Drexel University, Philadelphia 19104
+ Eastern College, St Davids 19087
+ LaSalle College, Philadelphia 19141
+ Rutgers-Camden, Camden NJ 08102
+ Temple University, Philadelphia 19122
+ Thomas Jefferson University, Philadelphia 19107
+ University of Pennsylvania, Philadelphia 19104
+ Villanova University, Villanova 19085
+ West Chester State College, West Chester 19380
+ Widener University, Chester 19013

752 Wilkes College, Wilkes-Barre 18766 (717) 829-0194
+ Bloomsburg State College, Bloomsburg 17815
+ College Misericordia, Dallas 18612
+ King's College, Wilkes-Barre 18711
+ Marywood College, Scranton 18509
+ The University of Scranton, Scranton 18510

PUERTO RICO

755 University of Puerto Rico, Rio Piedras, Rio Piedras 00936 (809) 767-1410
+ Bayamon Central University, Bayamon 00619
+ Bayamon Regional College, Rio Piedras 00931
+ Inter American University, Hato Rey 00919
+ Sacred Heart University, Santurce 00914
+ University of Puerto Rico, Bayamon Technical University College, Bayamon 00936
+ University of Puerto Rico, Cardina Regional College, Carolina 00931
+ University of Puerto Rico, Cayey University College, Cayey 00633
+ University of Puerto Rico, Humacao University College, Humacao 00661
+ World University, Hato Rey 00917

756 University of Puerto Rico, Mayaguez 00709 (809) 833-1888
+ Inter American University of Puerto Rico, San German 00753
+ University of Puerto Rico Aquadilla Regional College, Aquadilla 00603

SOUTH CAROLINA

765 The Citadel, Charleston 29409 (803) 792-5005
770 Clemson University, Clemson 29631 (803) 656-3254
+ Central Wesleyan College, Central 29630
772 Baptist College at Charleston, Charleston 29411
775 University of South Carolina, Columbia 29208 (803) 774-4134
+ Benedict College, Columbia 29204

SOUTH DAKOTA

780 South Dakota State University, Brookings 57007 (605) 688-6106

TENNESSEE

785 Memphis State University, Memphis 38152 (901) 454-2681
+ Christian Brothers College, Memphis 38104
+ LeMoyne-Owen College, Memphis 38126
+ Southwestern Memphis, Memphis 38112
+ University of Tennessee Medical School, Memphis 38163
790 Tennessee State University, Nashville 37203 (615) 320-3710
+ Belmont College, Nashville 37203
+ David Lipscomb College, Nashville 37203
+ Fisk University, Nashville 37203
+ Meharry Medical College, Nashville 37208
+ Middle Tennessee State University, Murfreesboro 37130
+ Trevecca Nazarene College, Nashville 37210
+ Vanderbilt University, Nashville 37240
+ Western Kentucky University, Bowling Green 42101
800 University of Tennessee, Knoxville 37996
+ Knoxville College, Knoxville 37921

TEXAS

805 Texas A&M University, College Station 77841 (409) 845-7611
810 Baylor University, Waco 76798 (817) 755-3513
+ University of Mary Hardin-Baylor, Belton 76513
+ Paul Quinn College, Waco 76704
820 Texas Tech University, Lubbock 79409 (806) 742-2143
+ Lubbock Christian College, Lubbock 79407
825 The University of Texas at Austin, Austin 78712 (512) 471-1776
+ Concordia Lutheran College, Austin 78705
+ St Edward's University, Austin 78704
830 East Texas State University, Commerce 75428 (214) 886-5200
835 North Texas State University, Denton 76203 (817) 565-2200
+ Southern Methodist University, Dallas 75275
+ Texas Woman's University, Denton 76204
+ University of Dallas, Irving 75061
840 Southwest Texas State University, San Màrcos 78666 (512) 392-8188
+ Texas Lutheran College, Seguin 78155
840A University of Texas at San Antonio, San Antonio 78285
845 Texas Christian University, Fort Worth 76129 (817) 921-7461

+ Texas Wesleyan College, Fort Worth 76105
+ University of Texas at Arlington, Arlington 76019

847 Angelo State University, San Angelo 76909 (915) 942-2036
+ University of Texas at El Paso, El Paso 79968

UTAH

850 University of Utah, Salt Lake City 84112 (801) 581-6236
+ Weber State College, Ogden 84408
+ Westminster College, Salt Lake City 84105

855 Brigham Young University, Provo 84602 (801) 378-2671

860 Utah State University, Logan 84322 (801) 750-1831

VERMONT

865 St Michael's College, Winooski 05404 (802) 655-2000 ext 2553
+ Lyndon State College, Lyndonville 05851
+ Trinity College, Burlington 05401
+ University of Vermont & State Agricultural College, Burlington 05401

867 Norwich University, Northfield 05663 (802) 485-5001 ext 269

VIRGINIA

875 Virginia Polytechnic Institute, Blacksburg 24060 (703) 961-6404

880 Virginia Military Institute, Lexington 24450 (703) 463-6354

890 University of Virginia, Charlottesville 22903 (804) 924-3394
+ George Mason University, Fairfax 22030

WASHINGTON

900 University of Puget Sound, Tacoma 98416 (206) 756-3264
+ Pacific Lutheran University, Tacoma 98447
+ St Martin's College, Olympia 98503
+ Southern Illinois University of McChord AFB 98438

905 Washington State University, Pullman 99164-2606 (509) 335-3546
+ University of Idaho, Moscow, Idaho 83843

910 University of Washington, Seattle 98195 (206) 543-2360
+ Seattle University, Seattle 98122

WEST VIRGINIA

915 West Virginia University, Morgantown 26506 (304) 293-5421
+ Fairmont State College, Fairmont 26554
+ Shepherd College, Shepherdstown 25443

WISCONSIN

925 University of Wisconsin, Madison 53706 (608) 262-3440
+ University of Wisconsin at Superior, Superior 54880

WYOMING

940 University of Wyoming, Laramie 82071 (307) 766-2338

REGIONAL ADMISSIONS COUNSELORS (ADCOs)

Regional Admissions Counselors (ADCOs) are located throughout the United States to provide information concerning Air Force college opportunities. (At times they cover areas in bordering states.) For specific information contact the ADCO or detachment nearest your location or call (205) 293-2091. Residents of Hawaii should contact the nearest detachment.

Alabama/Puerto Rico
Admissions Counselor
AFROTC Det 010
University of Alabama
University AL 35486
(205) 348-5900

Arizona/New Mexico
Admissions Counselor
AFROTC Det 025
Arizona State University
Tempe AZ 85287
(602) 965-3181

California/Nevada
Admissions Counselor
AFROTC/CC-WE
Norton AFB CA 92409
(714) 382-7801

Admissions Counselor
AFROTC Det 085
University of California, Berkeley
Berkeley CA 94720
(415) 642-3572

Admissions Counselor
AFROTC OL 055
University of California
Los Angeles CA 90024
(213) 825-1742

Florida
Admissions Counselor
AFROTC Det 159
University of Central Florida
Orlando FL 32816
(305) 275-2264

Georgia
Admissions Counselor
AFROTC Det 165
Georgia Institute of Technology
Atlanta GA 30332
(404) 894-4175

Hawaii
AFROTC Det 175
University of Hawaii
Honolulu HI 96822
(808) 948-7762

Illinois/Indiana
Admissions Counselor
AFROTC Det 195
Illinois Institute of Technology
Chicago IL 60616
(312) 567-3525

Illinois/Missouri
Admissions Counselor
AFROTC Det 207
Parks College of St Louis University
Cahokia IL 62206
(618) 337-7500 ext 230

Kansas/Missouri
Admissions Counselor
AFROTC Det 280
University of Kansas
Lawrence KS 66045
(913) 864-4676

Kentucky/Indiana
Admissions Counselor
AFROTC Det 290
University of Kentucky
Lexington KY 40506
(606) 257-1681

Louisiana/Texas
Admissions Counselor
AFROTC Det 310
Louisiana State Univ & A&M College
Baton Rouge LA 70893
(504) 388-4407

Maryland/DC/Delaware/Virginia
Admissions Counselor
AFROTC Det 330
University of Maryland
College Park MD 20742
(301) 454-3245

Massachusetts/Connecticut/Rhode Island
Admissions Counselor
AFROTC Det 370
University of Massachusetts
Amherst MA 01003
(413) 545-2437

Admissions Counselor
Det 115
University of Connecticut
Storrs CT 06268
(203) 486-2225

Michigan
Admissions Counselor
AFROTC Det 390
University of Michigan
Ann Arbor MI 48109
(313) 764-2405

Minnesota/ND/SD
Admissions Counselor
AFROTC Det 415
University of Minnesota
Minneapolis MN 55455
(612) 373-2205

Mississippi/Alabama
Admissions Counselor
AFROTC Det 432
University of Southern Mississippi
Hattiesburg MS 39401
(601) 266-4472

Nebraska/Iowa
Admissions Counselor
AFROTC Det 470
University of Nebraska at Omaha
Omaha NE 68182
(402) 554-2318

New Hampshire/Maine/Massachusetts
Admissions Counselor
AFROTC Det 475
University of New Hampshire
Durham NH 03824
(603) 862-1480

New Jersey/Delaware
Admissions Counselor
AFROTC Det 485
Rutgers—The State University
New Brunswick NJ 08901
(201) 932-7430

New Mexico/Texas
Admissions Counselor
AFROTC Det 505
New Mexico State University
Las Cruces NM 88003
(505) 646-2136

New York
Admissions Counselor
AFROTC Det 560
Manhattan College
Riverdale NY 10471
(212) 920-0203

Admissions Counselor
AFROTC Det 535
Syracuse University
Syracuse NY 13210
(315) 476-9272

North Carolina
Admissions Counselor
AFROTC Det 595
North Carolina State University
Raleigh NC 27650
(919) 737-2417

Ohio
Admissions Counselor
AFROTC Det 660
University of Akron
Akron OH 44325
(216) 375-7653

Ohio/West Virginia
Admissions Counselor
AFROTC Det 643
Wright State University
Dayton OH 45435
(513) 873-2730

Oklahoma/Arkansas
Admissions Counselor
AFROTC Det 675
The University of Oklahoma
Norman OK 73019
(405) 325-3211

Pennsylvania
Admissions Counselor
AFROTC Det 715
Lehigh University
Bethlehem PA 18015
(215) 861-3290

Admissions Counselor
AFROTC Det 730
University of Pittsburgh
Pittsburgh PA 15260
(412) 624-6397

South Carolina/Georgia
Admissions Counselor
AFROTC Det 775
University of South Carolina
Columbia SC 29208
(803) 777-4134

Tennessee
Admissions Counselor
AFROTC Det 790
Tennessee State University
Nashville TN 37203
(513) 873-2730

Texas
Admissions Counselor
AFROTC Det 840
Southwest Texas State University
San Marcos TX 78666
(512) 245-2182

Admissions Counselor
AFROTC Det 845
Texas Christian University
Fort Worth TX 76129
(817) 921-7461

Virginia
Admissions Counselor
AFROTC Det 890
University of Virginia
Charlottesville VA 22903
(804) 924-3394

Washington/Oregon/Alaska
Admissions Counselor
AFROTC Det 900
University of Puget Sound
Tacoma WA 98416
(206) 756-3264

Wisconsin
Admissions Counselor
AFROTC Det 925
University of Wisconsin-Madison
Madison WI 53706
(608) 262-3440

Wyoming/Idaho/Utah/Montana
Admissions Counselor
AFROTC Det 855
Brigham Young University
Provo UT 84602
(801) 378-2671

Vermont/New York
Admissions Counselor
AFROTC Det 550
Rensselaer Polytechnic Institute
Troy NY 12181
(518) 270-6236

Air Force Occupations

Career Fields	Duties & Responsibilities	Qualifications	Examples of Civilian Jobs
Accounting, Finance and Auditing	Prepares documents required to account for and disburse funds, including budgeting allocation, disbursing, auditing and preparing cost analysis records.	Dexterity in the operation of business machines. Typing, mathematics, statistics and accounting desirable. High administrative aptitude mandatory.	Public accountant, auditor, bookkeeper, budget clerk and paymaster.
Administrative	Prepares correspondence, statistical summaries, arranges priority and distribution systems, maintains files, prepares and arranges for graphic presentation.	English, typing, accounting, mathematics and shorthand courses desirable.	Clerk typist, file clerk, secretary, stenographer, receptionist.
Aircraft Maintenance	Performs the mechanical functions of maintenance, repair and modification of helicopters, turboprop, reciprocating engine and jet aircraft.	Considerable mechanical or electrical aptitude and manual dexterity. Shop mathematics and physics desirable.	Aircraft mechanic, airframe inspector.
Aircraft Systems Maintenance	Performs maintenance of aircraft accessory systems, propulsion systems, fabrication of metal and fabric materials used in aircraft structural repair, and inspection and preservation of aircraft parts and materials.	Electrical and mechanical aptitude; shop mathematics and shopwork is desirable.	Aircraft mechanic, aircraft electrician, sheet metal worker, welder, machinist.
Aircrew Protection	Teaches the use of survival techniques and protective equipment.	Good physical condition required, knowledge of pioneering and woodsman activities helpful.	No civilian job covers the scope of the jobs in this career field, but a related job is that of hunting or fishing guide.

Career Fields	Duties & Responsibilities	Qualifications	Examples of Civilian Jobs
Audiovisual	Operates aerial and ground cameras, motion picture and other photographic equipment; processes photographs and film, edits motion pictures, performs photographic instrumentation functions, and operates airborne, field and precision processing laboratories.	Considerable dexterity on small precision equipment; excellent eyesight. Mathematics, physics and chemistry desirable.	Photographer, darkroom technician, film editor, aerial commercial photographer, photograph finisher, sound mixer and motion picture camera operator.
Avionics Systems	Installs, maintains and repairs airborne bomb navigation, fire control, weapon control, automatic flight control systems, radio, and navigation equipment and maintains associated test equipment.	Electronic aptitude, manual dexterity and normal vision. Mathematics, physics desirable.	Radar, television and precision instrument maintenance.
Band	Plays musical instruments in concert bands and orchestras, repairs and maintains instruments, vocalist, performs as drum major, arranges music and maintains music libraries.	Knowledge of rudiments of music, elementary theory of music and orchestration desirable.	Orchestrator, music librarian, music teacher, instrumental musician.

Communication-Electronic Systems	Installs, modifies, maintains repairs and overhauls airborne and ground television equipment, high speed general and special purpose data processing equipment, automatic communications and cryptographic machines systems, teletypewriter, teleautographic equipment, telecommunications systems control and associated electronic test equipment.	Basic knowledge of electronic theory. Mathematics and physics desirable. Normal color vision mandatory.	Communications, electronics technician, radio and television repairer, meteorological and teletype equipment repairer.
Command Control Systems Operations	Operates control towers, directs aircraft landings with radar landing control equipment; operates ground radar equipment, aircraft control centers, airborne radar equipment, space tracking and missile warning systems.	Equipment dexterity, clear voice and speech ability and excellent vision. English desirable.	Aircraft log clerk, airport control operator and air traffic controller.
Communications Operations	Operates radio and wire communications systems, automatic digital switching equipment, cryptographic devices, airborne and ground electronic countermeasures equipment, all kinds of communication equipment, and the management of radio frequencies.	Knowledge of telecommunications functions and operations of electronic communications equipment. Typing and clear speaking voice desirable in many specialties.	Central office operator (telephone and telegraph), cryptographer, radio operator, telephone supervisor and photo-radio operator.

Career Fields	Duties & Responsibilities	Qualifications	Examples of Civilian Jobs
Computer Systems	Collects, processes, records, prepares and submits data for various automated systems, analyzes design, programs and operates computer systems.	Business math, algebra, geometry desirable.	Card-tape converter or computer operator, data typist, data processing control clerk, high-speed printer operator, programmer.
Contracting/ Manufacturing	Acquires material and services for support of military installations; involves advertising or negotiating bids and awarding contracts.	General aptitude for business and clerical work; business arithmetic and bookkeeping courses desirable.	Purchasing agent, contract negotiator, contract specialist, procurement and general clerk.
Dental	Operates dental facilities and provides paraprofessional dental care; preventive dental services, treatment of oral tissues and fabricates prosthetic devices.	Knowledge of oral and dental anatomy; physiology, biology and chemistry desirable.	Dental hygienist, dental assistant.
Education and Training	Conducts formal classes of instruction, uses training aids, develops material for various courses of instruction, teaches classes in general academic subjects and military matters, and administers educational programs.	English composition and speech desirable.	Vocational training instructor, counselor, educational consultant, or administrator.

Enlisted Aircrew	Primary duties require frequent and regular flights. Inflight Refueling Operator performs duties associated with inflight refueling of aircraft; Aircraft Loadmaster supervises loading of cargo and passengers and operates aircraft equipment; and Flight Engineers ensure mechanical condition of the aircraft and monitor inflight aircraft systems.	High electrical and mechanical aptitude, manual dexterity, normal vision and good physical condition. Mathematics, physics and shopwork desirable.	Aircraft mechanic, electrician, hydraulic tester, oxygen systems tester, cargo handler, dispatcher and shipping clerk depending upon the area in which training and experience is received. No civilian job covers some aspects of this field.
Fire Protection	Operates firefighting equipment, prevents and extinguishes aircraft and structural fires; rescues and renders first aid; maintains firefighting and fire prevention equipment.	Good physical condition; no allergies to oil and fire extinguishing solutions; general science and chemistry desirable.	Fire chief, fire extinguisher serviceperson, firefighter, fire marshal and fire department equipment person.
Fuels	Receives, stores, dispenses, tests and inspects propellants, petroleum fuels and products.	Chemistry and math.	Petroleum industry supervisor and bulk plant manager.

Career Fields	Duties & Responsibilities	Qualifications	Examples of Civilian Jobs
Intelligence	Collects, produces and disseminates data of strategic, tactical or technical value from an intelligence viewpoint. Maintains information security.	Knowledge of techniques of evaluation, analysis, interpretation and reporting, foreign languages, English composition, photography and mathematics desirable.	Cryptoanalyst, draftsperson, interpreter, investigator, statistician, radio operator and translator.
Intricate Equipment Maintenance	Overhauls and modifies photographic equipment; works with fine precision tools, testing devices and schematic drawings.	Considerable mechanical ability and manual dexterity; algebra and physics desirable.	Camera repairer, statistical machine and medical equipment serviceperson.
Legal	Takes and transcribes verbal recordings of legal proceedings; uses stenomask; performs office administrative tasks; processes claims.	Knowledge of stenomask, typewriter, legal terminology, military legal procedures, preparation and processing of claims; English grammar and composition; ability to take dictation by stenomask at 175 words per minute, type 50 words per minute and speak clearly and distinctly.	Law librarian, court clerk and shorthand reporter.
Management Analysis	Collects, processes, records, controls, analyzes, and interprets special and recurring reports, statistical data and other information.	Knowledge of business statistics, mathematics, accounting and English desirable. Completion of high school or GED equivalent mandatory.	Statistical, accounting and budget clerk.
Manpower Management	Performs management studies and evaluates work requirements to determine manpower required to accomplish Air Force jobs. Conducts management consultant studies.	Completition of high school or GED equivalency with courses in mathematics, including algebra and high general aptitude mandatory.	Industrial engineering technician, management and job analyst.

Marine	Performs duties in the operation and maintenance of boats, deck equipment, gasoline or diesel engines; prepares boats and related equipment for storage; uses utility craft, inspects and repairs mechanical, electrical, and sanitary marine equipment and maintains boat structure.	Mechanical aptitude; normal color vision and high vision; normal depth perception, hearing, and gait and balance; no record of acrophobia and physical ability to perform climbing duties; no record of hydrophobia.	Ordinary seaman, motorboat operator, ship electrician, marine engine machinist, motorboat mechanic, marine oiler.
Mechanical/ Electrical	Performs installation, operation, maintenance and repairs of base direct support systems and equipment.	Physics, mathematics, blueprint reading and electricity.	Elevator repairer, electrician, powerhouse repairer, diesel mechanic, pipefitter, steamfitter and heating and ventilating worker.
Medical	Operates medical facilities, works with professional medical staff as they provide care and treatment. May specialize in such medical services as nuclear medicine, cardiopulmonary techniques, physical and occupational therapy, orthopedic appliances, medical laboratory, veterinary and medical administrative services.	Knowledge of first aid, ability to help professional medical personnel; anatomy, biology, zoology. High school algebra and chemistry desirable in most specialties and are mandatory requirements for some.	X-ray and medical record technician, medical laboratory and pharmacist assistant. Respiratory therapy technician and surgical technologist.

Career Fields	Duties & Responsibilities	Qualifications	Examples of Civilian Jobs
Missile Electronics Maintenance	Assembles, installs, maintains, checks out, repairs, and modifies missile and drone guidance, control, analyst, launches, test equipment, and instrumentation systems.	Knowledge of basic electronic theory and circuits, hydraulics. Normal color vision mandatory.	Missile facilities repairer, electronics engineer, electronics inspector, electronics mechanic, and aircraft mechanical electrical system repairer.
Missile Maintenance	Performs missile engine installation, maintenance and repair; maintenance, repair and modification of missile airframes, subsystems and associated aerospace ground equipment.	Mechanical aptitude and manual dexterity. Mathematics and physics desirable. Normal color vision mandatory.	Mechanical inspector, mechanical engineer, aircraft mechanic, and pneumatic tester and mechanic.
Moral, Welfare and Recreation	Conducts physical conditioning, coaches sports program, administers recreation, entertainment, sports and club activities.	Good muscular coordination; English, business math and physical education desirable.	Athletic or playground director, physical instructor and manager of a recreational establishment.
Motor Vehicle Maintenance	Overhauls and maintains powered ground vehicles and mechanical equipment for transporting personnel and supplies.	Machine shop, mathematics and training in the use of tools and blueprints helpful.	Automobile accessories installer, automobile repairer, bus mechanic, automotive electrician and truck mechanic.
Munitions and Weapons Maintenance	Maintains and repairs aircraft armament, assembles, maintains, loads, unloads and stores munitions and nuclear weapons; disposes of bombs, missiles and rockets and operates detection instruments.	Mechanical or electrical aptitude, manual dexterity, normal color vision and depth perception; mathematics and mechanics desirable.	Aircraft armament mechanic. armorer, ammunition inspector, munitions handler.

Personnel	Interviews, classifies, selects career jobs for airman on the basis of qualifications and requirements of the Air Force; administers aptitude, performance tests; administers personnel quality control programs; performs counseling, educational and administrative functions.	English composition, speech and social science. Operation of simple data processing equipment and typing ability desirable.	Employment or personnel clerk, special services supervisor, personnel service manager, personnel supervisor, counselor.
Photomapping	Procures, compiles, computes and uses topographic, photogrammetric, and cartographic data in preparing aeronautical charts, topographic maps and target folders.	Ability to use precision instruments required in measuring and drafting; algebra, geometry, trigonometry and physics necessary.	Cartographer, topographical draftsperson, mapmaker, and advertising layout person.
Printing	Operates and maintains reproduction equipment used in the graphic arts, performs hand and machine composition and binding operations.	Mechanical ability and dexterity; courses in chemistry and shop mechanics desirable.	Lithographic press, fold machine, offset and webpress, perforating machine and duplicating machine operator; bookbinder; photolithographer; photoengraver.
Public Affairs	Interviews people; reports news; composes, proofreads, writes and edits news copy; provides public affairs advice.	High general aptitude and completion of high school or GED equivalency mandatory. English grammar and composition, speech, journalism, drama, radio/television, history, or political science desirable.	Reporter, copy reader, historian, public relations representative, editorial assistant, broadcast or program director, announcer.

Career Fields	Duties & Responsibilities	Qualifications	Examples of Civilian Jobs
Safety	Conducts safety programs, surveys areas and activities to eliminate hazards, analyzes accident causes and trends.	Knowledge of industrial hygiene, safety education, safety psychology, and blueprint interpretation; typing, English and public speaking desirable.	Safety inspector and instructor.
Sanitation	Operates and maintains water and waste processing plants' systems and equipment and performs pest and rodent control functions.	Physics, biology, chemistry and blueprint reading valuable.	Purification plant operator, sanitary inspector, exterminator and entomologist.
Security Police	Provide security and classified information and material, enforce law and order, control traffic, and protect lives and property, organize as local ground defense forces.	Good physical condition, vision and hearing; civics and social sciences desirable.	Guard, police inspector, police officer and superintendent of police.
Special Investigations/ Counter-intelligence	Investigates violations of the Uniform Code of Military Justice and applicable federal statutes, investigates conditions pertaining to sabotage, espionage, treason, sedition and security.	Knowledge of law enforcement and security regulations, good physical condition, hearing and vision; civics, social sciences, accounting and foreign language desirable.	Detective, chief of detectives, detective sergeant and investigator.
Structural/ Pavements	Constructs and maintains structural facilities and pavement area; maintains pavements, railroads and soil bases; performs erosion control; operates heavy equipment; performs site development, general maintenance, cost and real property accounting, work control functions and metal fabricating.	Blueprint reading, mechanical drawing, mathematics, physics, and chemistry.	Plumber, bricklayer, carpenter, stonemason, painter, construction worker, welder and sheet metal worker.

Supply	Designs, develops, analyzes and operates supply systems including supply data systems; responsible for computation, operation and management of material facilities; equipment review and validation; records maintenance, inventory and distribution control; budget computation and financial plans.	Accounting and business administration.	Junior accountant, machine records section supervisor, receiving, shipping and stock clerk.
Supply Service	Supervises and operates sales stores, laundry/dry cleaning facilities, commissaries and meat processing. Cooks and bakes.	Chemistry, management, marketing, manual dexterity and business mathematics.	Department manager, retail general merchandise manager, meat cutter, butcher, chef and pastry cook.
Training Devices	Installs, operates, repairs, and modifies instrument, navigation, bombing gunnery trainers and flight simulators; works with small tools and precision test equipment.	Knowledge of electricity, mathematics, blueprint reading and physics desirable.	Link trainer instructor, radio mechanic.
Transportation	Ensures service, efficiency and economy of transportation of supplies and personnel by aircraft, train, motor vehicle and ship.	Driver training, operation of office machines and business math.	Cargo handler, motor vehicle dispatcher, shipping or traffic rate clerk, trailer truck driver and ticket agent.

Career Fields	Duties & Responsibilities	Qualifications	Examples of Civilian Jobs
Wire Communications Systems Maintenance	Installs and maintains wire communications equipment and systems. Installs, repairs and maintains telephone and telegraph land line systems, telephone equipment, antenna support systems, key systems, telephone switching equipment, missile communications control systems and electronic switching equipment.	Mechanical aptitude and manual dexterity; physics and mathematics desirable. Normal color perception mandatory. Physical ability to climb required in some specialties.	Cable splicer, central office repairer, line installer and inspector, telephone inspector, teletype and central office manual equipment repairer.
Weather	Collects, records and analyzes meteorological data; makes visual and instrument weather observations. Forecast immediate and long-range weather conditions.	Visual acuity correctable to 20/20; physics, mathematics and geography desirable.	Meteorologist, weather forecaster and weather observer.

Chapter V

The United States Coast Guard

O beautiful for spacious skies,
For amber waves of grain,

For purple mountain majesties
Above the fruited plain!

America! America!
God shed His grace on thee

And crown thy good with brotherhood
From sea to shining sea!

—from AMERICA THE BEAUTIFUL
by Katharine Lee Bates

The Coast Guard has come a long way since 1790, when it was founded by Alexander Hamilton as the Revenue Cutter Service. In 1915 it received its present name, but actually Coast Guard is a misnomer. You might suppose its sole task was guarding our coasts in peacetime, but Coast Guard activities are much more exciting than that. The Coast Guard serves in war as well as peace, not just on our coasts but overseas, and not just at sea but in the air and even on land.

With the aid of modern technology, sonar and radar, the Coast Guard is instrumental in rescuing ships in distress and through the years has saved thousands of lives. Shipwrecks were the way of the past. Just recently treasure hunters salvaged the Spanish galleon *Muestra Senora de Atocha*. Victim of a 1622 hurricane, the proud ship went down with some four hundred million dollars worth of gold and silver bars and other treasure. A few years ago we read of a Grecian vessel of even earlier date yielding treasures from ancient times. Even as the Coast Guard has come a long way, so has weather reporting and nautical technology. No more Titanics!

If you decide to serve in the Coast Guard, you will have multiple choices, for the Coast Guard has multiple missions.

Search and Rescue. The Coast Guard's large fleet of the most seaworthy vessels in the world and its strategically placed air stations controlled by district centers are ever ready. The Coast Guard has saved billions of dollars worth of property as well as thousands of lives.

International Ice Patrol. The Coast Guard patrols some 45,000 square miles in the North Atlantic, spotting icebergs and warning the many ships that pass through there during the ice season from April to July. With air and radar, it is the Coast Guard's mission to assure that never again will there be such a catastrophe as the sinking of the Titanic in 1912, when many lives were lost.

Aids to Navigation. From time immemorial, the lighthouse has been a symbol of hope. In addition to lighthouses, the Coast Guard maintains more than 46,000 navigational aids—radio beacons, buoys, fog signals, and others—and electronic aids such as loran, which make possible the quick discernment of position of air and surface craft. All this prevents many shipwrecks and makes life at sea and in the air safer.

Marine Environment Protection. Long before environmental news was making headlines, the Coast Guard was at work guarding the quality of coastal waters and the oceans. As the years have passed, it has become of increasing concern to protect the environment, to remove oil slicks from coastal waters and enforce the regulatory laws.

U.S. COAST GUARD PHOTO

A musician plays with the Coast Guard Band.

Port Safety. This vital mission is assigned exclusively to the Coast Guard. The Captain of the Port monitors traffic to prevent collisions and groundings in the port and harbor that is his responsibility.

Recreational Boating Safety. The ten Coast Guard districts maintain Boating Safety Offices, which direct Boating Safety Detachments patrolling territorial waters to save lives and enforce the law.

Enforcement of Laws and Treaties. The Coast Guard works in close cooperation with the Customs and Narcotics bureaus, the Department of Justice, and other government agencies to administer the many laws that protect the interests of the United States on waters subject to its jurisdiction and on the high seas. Such matters include who shall take the fish and which species should be protected, rules for invasion of coastal waters by foreign vessels, the proper shipment and discharge of merchant vessel crews, and the development and promulgation of new or revised standards and rules for marine safety.

Merchant Marine Safety. This primarily involves the seaworthiness of merchant vessels. Enforcement of the navigation laws of the United States are a responsibility of the Coast Guard on a national and international basis.

Icebreaking. The Coast Guard operates all U.S. icebreakers, cutters that make regular cruises to the Arctic and Antarctic.

Defense Readiness. While continuously busy in peacetime in lifesaving and patrol missions, the Coast Guard is also alert in times of war, guarding convoys, accomplishing rescue missions, and conducting antisubmarine patrols and transport and combat landing craft operations.

Organization of the Coast Guard

At all times the Coast Guard is a military service and one of the five branches of the United States armed forces. It operates as a service under the Department of Transportation in times of peace and a part of the Navy in times of war.

The Commandant, assisted by the Headquarters staff, plans, directs, coordinates, and evaluates Coast Guard activities carried out by the Atlantic and Pacific Area Commanders and provides immediate direction to units assigned directly to headquarters.

The Area Commanders, assisted by their staffs, provide direction, support, and coordination of specified operational and support functions that involve the activities of more than one District.

The District Commanders, assisted by their staffs, provide regional direction, support, and coordination for functions performed by subordinate units assigned.

Activities, Section, and Group Commanders, assisted by their staffs, provide direction, support, and coordination for functions performed by subordinate units assigned.

Field Units execute and support the missions, programs, and functions assigned to the Coast Guard.

Headquarters Units provide support services for the Coast Guard as a whole and are under the immediate direction of the Commandant, assisted by the Headquarters Staff.

Coastal defense has become increasingly important in U.S. defense planning, and in 1984 the Coast Guard was assigned a new function in that area with establishment of the Maritime Defense Zone Pacific (MarDeZPac). A joint command with the U.S. Navy, it is generally concerned with waters within the exclusive economic zone; planning focuses on the 200 nautical miles adjacent to the coast. Sector commanders are either Navy or Coast Guard officers. Coast defense planners and surveillance systems of both services work side by side.

The United States Coast Guard Academy

As the Coast Guard is the smallest of the armed forces, its officers receive command responsibility early in their career, and they must be well trained and dedicated to uphold the traditions of the service. Such is the purpose of the Coast Guard Academy, situated on a 100-acre site on the west bank of the Thames River in New London, Connecticut. With Coast Guard ships in a nautical environment, the Academy has scientific, computer, engineering, library, gymnasium, and residence facilities that would be hard to surpass.

Admissions Requirements

Unlike the other service academies, the Coast Guard has no Congressional appointments or geographical quotas. Candidates are selected on the basis of an annual nationwide competition. Criteria are high school rank, performance on either the College Board Scholastic Aptitude Test (SAT) or the American College Testing Program (ACT), plus leadership potential as evidenced by high school extracurricular participation, community projects, or part-time employment.

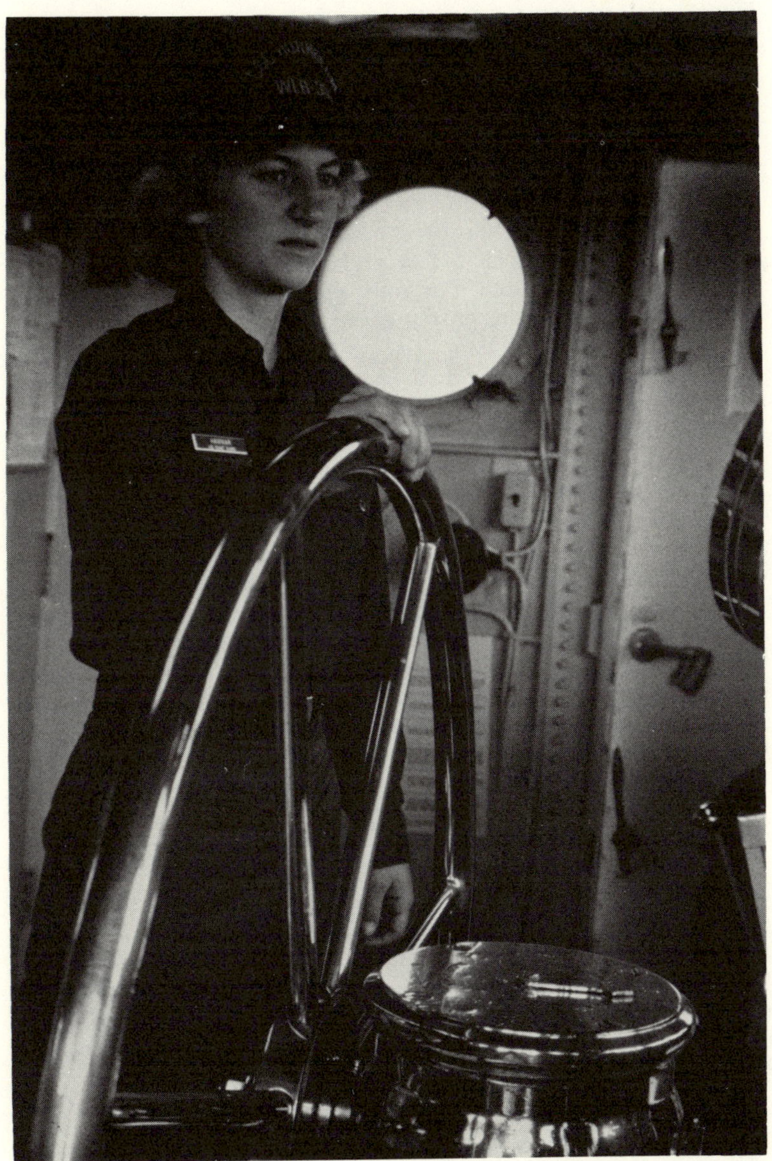

U.S. COAST GUARD PHOTO

A Coast Guardswoman takes the helm.

Eligibility Requirements

You must have reached the age of seventeen but not the age of twenty-two by July 1 of the year you are to be admitted as a cadet. You must be a U.S. citizen at the time of entry into the academy. You must be unmarried at the time of appointment and have no legal obligations as a result of a prior marriage. A cadet may not marry until after graduation from the Academy. You must have good moral character, trustworthiness, and emotional stability.

You must be between 5 feet and 6 feet 6 inches, with weight suitable to your physique. You must pass a Service Academy Medical Examination before entering the Academy and show strong physical aptitude.

Scholastically, you must be a graduate (or senior assured of graduation) of an accredited high school or preparatory school by June 30 of the year of entrance into the Academy.

The transcript you submit must show a total of 15 units, 6 of which are mandatory—3 of mathematics (algebra, plane or coordinate geometry, or equivalents), and English, 1, 2, and 3. Seven optional units are required from physical science (physics, chemistry, general science, etc.); additional English; additional math; social science (history, civics etc.); biological science (biology, zoology, etc.), and not less than 2 units in a foreign language. It is strongly recommended that you include solid geometry, trigonometry, physics, and chemistry in your preparation for a Coast Guard career.

When appointed a cadet, you agree to serve five years in the U.S. Coast Guard after graduation from the Academy.

Women at the Coast Guard Academy

Women undergo basically the same training as their male counterparts, with slightly modified physical training. The academic curriculum is the same for both men and women, as are the leadership and professional courses. Women are fully integrated into the cadet barracks but have women roommates and separate showers, toilets, and sinks. Their hair is worn short, off the collar for easy care. Cosmetics and jewelry are limited to essentials (minimal makeup, wristwatches, and post earrings for pierced ears).

Academic Program

The Academic Division consists of eight departments under the direction and supervision of the Dean of Academics: Engineering,

Nautical Science and Law, Mathematics, Science, Economics and Management, Computer Science, Humanities, and Physical Education. Currently the following majors are offered: Marine Engineering, Electrical Engineering, Civil Engineering, Marine Sciences, Mathematical and Computer Science, Management and Government. The Marine Sciences major combines the studies of ocean engineering and marine and physical sciences. It provides a strong foundation in such diverse areas as marine pollution and physical and chemical aspects of the ocean, as well as solid engineering subjects connected with ocean systems.

Funds

Each cadet receives a basic allowance of $590.00 per month plus a subsistence allowance of $59.53 a month. This is not a wage or salary, but is money furnished by the government for uniforms, equipment, textbooks, and other expenses incidental to training; it is disbursed and expended only as directed by the Superintendent of the Academy. Any funds remaining are given to the cadet upon graduation. For personal use the cadet receives a monthly allowance as follows: $80 for first year; $150 for second year; $210 for third year; $260 for fourth year.

Coast Guard Officer Candidate School

Selected college graduates are trained for commissioning in a course at Officer Candidate School, Yorktown, Virginia. A candidate must be a U.S. citizen and a graduate of an accredited college or university. Consistent with the needs of the Coast Guard, duty assignments are made in consideration of educational background, experience, and preference. Age limits are twenty-one through twenty-six as of class convening date.

Selected candidates are enlisted in the Coast Guard Reserve and assigned to inactive duty while awaiting the convening date of the next class. They may follow their usual civilian pursuits during this time.

Classes are 17 weeks in duration. You remain in enlisted status until graduation and are paid every two weeks at a base pay of about $500 a month. While at OCS you may indicate the type of duty and location you prefer after graduation. Your wishes will be taken into consideration along with the needs of the service and your experience and educational background.

Graduates are commissioned as ensigns and are obligated to serve on active duty for three years. Those who wish may apply for integration into the Regular Coast Guard and, if accepted, continue to serve as career officers.

Officer Promotion

The Coast Guard operates on a General Duty concept for its officers. Approximately 50 percent of them are deck officers who serve about half their time at sea. The other 50 percent who have taken advantage of specialized training opportunities serve tours of duty in the field of their choice, dictated by the needs of the Coast Guard, and may serve alternately at sea if they keep up their qualifications. The objective is to produce a well-versed officer with a well-rounded career pattern.

There are ample opportunities for sea duty for most female serious members who desire jobs afloat. With the exception of an Academy graduate's first assignment, sea duty is not mandatory for career or promotion.

An ensign is promoted to lieutenant (j.g. or junior grade) after approximately eighteen months of active duty, to lieutenant after approximately five years of commissioned service, to lieutenant commander after approximately eleven years of commissioned service; and to commander after about fifteen years of commissioned service. Promotions are based on fitness reports, which are given twice yearly for every officer. A selection board, which meets once a year, nominates the best-qualified officers for promotion.

Direct Commissions

Graduates of a law school accredited by the American Bar Association may receive commissions as lieutenants in the Coast Guard Reserve. Application to take the bar exam or actual admission must be evidenced prior to commissioning, and the graduate must be admitted to the bar of a state or federal court within one year after being commissioned. The Coast Guard secures marine safety officers through its Maritime Academy Graduate program.

Aviation Programs

Coast Guard pilot training is available to selected graduates of Officer Candidate School or the Coast Guard Academy. Pilot trainees attend fourteen months of basic and advanced flight training at naval air stations in Pensacola, Florida, and Corpus Christi, Texas, or at Mather AFB, Sacramento, California. Applicants must not have reached their twenty-eighth birthday at commencement of flight training.

Medical Programs

Students enrolled in an accredited medical or osteopathy school may receive $600 per month, plus tuition and fees. Upon receipt of their degree, they serve with the U.S. Public Health Service, on active duty with the Coast Guard or other USPHS agency for one year for every year of subsidized training or a minimum of two years. To apply, contact the National Health Service Corps, Scholarship Programs, Center Building, Room 5-44, 3700 East-West Hwy., Hyattsville, Md. 20782.

Enlistment in the Coast Guard

If you enlist in the Coast Guard, you can expect eight weeks of basic training at Camp May, New Jersey. In addition to courses basic to all services, Coast Guard trainees study seamanship, ordnance (military weapons, ammunition, etc.), damage control, and Coast Guard history and traditions. After that most graduates receive technical training by attending schools or through a carefully planned on-the-job program.

Promotion

A Coast Guard seaman recruit (E-1) is promoted to seaman apprentice (E-2) upon completion of basic training. Eligibility for promotion to seaman or fireman (E-3) is based on five requirements: adequate time in grade, successful demonstration of military and professional qualifications, passing of local examinations, recommendation of the commanding officer, and completion of required correspondence courses.

The Coast Guard provides four ways to advance to petty officer depending on current procedures and regulations:

1. Enlist under the guaranteed school program and have a written guarantee of assignment to a specific basic petty officer school before you go to recruit training.
2. Decide upon the specific petty officer school while on recruit training and qualify for it then.
3. Wait until after recruit training to get a more general knowl-

edge of the Coast Guard and then decide upon the right school.

4. Aim for petty officer status by actually performing on-the-job duties, completing correspondence courses in your chosen specialty, and passing a written examination.

Further promotions are on a service-wide basis, which gives you more chance for the next higher grade as opportunities in the entire service open up.

To go from enlisted to warrant officer, you must have earned this special status by ability and experience. The warrant officer is a technical specialist.

From Enlisted to Officer. If interested and qualified, you will be encouraged to apply for Officer Candidate School as you become eligible. If selected, you go to the Coast Guard Reserve Training Center in Yorktown, Virginia, for seventeen weeks of intensive instruction in navigation, seamanship, communications, and leadership. Upon graduation, you are commissioned an ensign.

Educational Opportunities

The Coast Guard sponsors a tuition assistance program of off-duty education within the limits of available funds. It is designed to increase the technical and educational background of Coast Guard personnel and to assist in the attainment of personal education goals.

Regular Coast Guard and Coast Guard Reserve personnel on active duty are eligible to participate in the program, which pays up to 100 percent of tuition fees.

Coast Guard Reserve

Qualified persons may enter the Coast Guard Reserve through one of three programs. First, candidates without prior service, seventeen through twenty-five years old, enlist for six years. This involves active duty for basic training and specialized schooling for enlistees without service-related skills.

Second, in a special program for students, the initial active duty period is split between two summers. Reservists then attend meetings one weekend every month and two weeks of active duty annually for the remainder of the eight-year enlistment.

Third, men and women twenty-six through thirty-four years of age having a needed skill may enlist for three years and receive a petty officer rating. This requires two weeks of orientation at Yorktown, Virginia, or Alameda, California. Those enlisting with service-related skills may be given the pay grade E-4. (Prior-service men and women up to age forty-two who were E-3 and above are eligible to reenlist. Those enlisting with prior service may be given rates corresponding to their civilian experience up to E-6.)

After training or orientation, Reservists return to their local units and serve one weekend a month plus two weeks annual active duty.

Coast Guard Occupations

Career Fields	Duties & Responsibilities	Qualifications	Examples of Civilian Jobs
Aviation Electrician's Mate	Maintains, adjusts, repairs aircraft electrical and instrument systems, plus power generating, lighting, electrical components of aircraft controls.	Algebra, trigonometry, physics and shop experience in aircraft electrical work.	Aircraft electrician; electrician, and TV repairer.
Aviation Electronics Technician	Tests, maintains, repairs aviation electronics equipment including navigation, identification, detection, reconnaissance, special purpose equipment; operates warfare equipment.	High degree of aptitude for electrical and mechanical work. Algebra, trigonometry, physics, electricity, radio and mechanics.	Aircraft electrician, radio mechanic, electrical repairer, instrument repairer, electronics technician, TV repairer.
Aviation Machinist's Mate	Inspects, maintains power plants and related systems and equipment, prepares aircraft for flight, conducts periodic aircraft inspections.	Good learning ability and mechanical aptitude. Machine shop, automobile or aircraft engine work, algebra and geometry.	Airport serviceperson, aircraft engine test mechanic, small appliance repairer.
Aviation Structural Mechanic	Maintains and repairs aircraft, airframe, structural components, hydraulic controls, utility systems, egress systems.	High degree of mechanical aptitude. Metal shop, woodworking, algebra, plane geometry, physics; experience in automobile body work.	Welder, sheet metal repairer, hydraulics technician, air conditioning repairer.

Aviation Survivalman	Maintains and packs parachutes, survival equipment, flight and protective clothing, life jackets; tests and services pressure suits, cares for search and rescue pyrotechnics and station small arms.	Must perform extremely careful and accurate work. General shop and sewing desirable. Experience in use and repair of sewing machines.	Parachute packer, inspector, repairer and tester; sailmaker.
Boatswain's Mate	Performs seamanship tasks, operates small boats, stores cargo, handles rope and lines, directs work of deck force personnel.	Must be physically strong, Practical math desirable; algebra, geometry and physics.	Motorboat operator, pier superintendent, able seaman, canvas worker, rigger, cargo wincher, mate, longshore worker.
Damage Controlman	Fabricates, installs, repairs shipboard structures, plumbing and piping systems, uses damage control in firefighting, and nuclear, biological, chemical and radiological defense equipment.	High mechanical aptitude. Sheet metal, foundry, pipefitting, carpentry mathematics, geometry and chemistry valuable.	Firefighter, welder, plumber, shipfitter, blacksmith, metallurgical technician.
Electrician's Mate	Maintains power and lighting equipment generators, motors, power distribution systems, other electrical equipment; rebuilds electrical equipment.	Aptitude for electrical and mechanical work. Electrical, practical and shop mathematics, and physics.	Electrician, electric motor and electrical equipment repairer, armature winder, radio/TV repairer.
Electronics Technician	Maintains all electronic equipment used for communications, detection ranging, recognition and countermeasures.	Aptitude for detailed mechanical work. Radio, electricity, physics, algebra, trigonometry and shop valuable.	Electronics technician, radar and radio repairer, instrument and electronics mechanic.

Career Fields	Duties & Responsibilities	Qualifications	Examples of Civilian Jobs
Fire Control Technician	Operates, tests, maintains and repairs weapons control systems and telemetering equipment used to compute and resolve factors which influence accuracy of naval guns and missiles.	Perform fine, detailed work. Extensive training in mathematics, mechanics.	Radar or electronics technician, test range tracker, instrument repairer, electrician.
Gunner's Mate	Operates and performs maintenance on guided-missile launching systems, rocket launchers, guns, gun mounts; inspects/repairs electrical, electronic, pneumatic, mechanical and hydraulic systems.	Prolonged attention and mental alertness, ability to perform detailed work. High aptitude for electrical and mechanical work. Arithmetic, shop math, electricity, electronics, physics, machine shop, welding, mechanical drawing and shopwork.	Electronics technician, electrician, instrument repairer, hydraulics, pneumatic or mechanical technician, small appliance or test equipment repairer.
Health Services	Administers medicines, applies first aid, assists in operating room, nurses sick and injured.	Hygiene, biology, first aid, physiology, chemistry, typing and public speaking.	Practical nurse, medical or X-ray lab technician, nurse administrator.
Machinery Technician	Operates, maintains and repairs ship's propulsion, auxiliary equipment and outside equipment such as steering engine, refrigeration/air conditioning, laundry equipment.	Aptitude for mechanical work. Practical or shop mathematics, machine shop, electricity and physics valuable.	Boiler house repairer, engine maintenance machinist, marine engineer, turbine operator, engine repairer, air conditioning and refrigeration repairer.
Marine Science Technician	Makes visual/instrumental weather and oceanographic observation; conducts chemical analysis; enters data on appropriate logs, charts, and forms; analyzes/interprets weather and sea conditions.	Ability to use numbers in practical problems. Algebra, geometry, trigonometry, physics, physiography, chemistry, typing, meteorology, astronomy and oceanography useful.	Oceanographic technician, weather observer, meteorologist, chart maker, statistical clerk and inspector of weather and oceanographic instruments.

Public Affairs Specialist	Reports, edits, copyreads news about service people and activities through newspapers, magazines, radio and television; takes still news photographs.	High degree of journalistic aptitude. English, journalism, typing, photography, and writing experience helpful.	News editor, copyreader, script writer, reporter, free lance writer, rewrite or art layout person, producer, photographer.
Quartermaster	Performs navigation of ships, steering, lookout supervision, ship-control, bridge watch duties, visual communication and maintenance of navigational aids.	Good vision and hearing and ability to express oneself clearly in writing and speaking. Public speaking, grammar, geometry and physics helpful.	Barge, motorboat, yacht captain, quartermaster, harbor pilot aboard merchant ship.
Radarman	Operates surveillance and search radar, electronic recognition and identification equipment, controlled approach devices and electronic aids to navigation; serves as plotter and status board keeper.	Prolonged attention and mental alertness. Physics, mathematics and ship courses in radio and electricity helpful. Experience in radio repair is valuable.	Radio operator (aircraft, ship, government service, radio broadcasting), radar equipment supervisor, and control tower operator.
Radioman	Operates communication, transmission, reception, and receives and processes all forms of military record and voice communications.	Good hearing and manual dexterity. Mathematics, physics and electricity desirable. Experience as amateur radio operator helpful.	Telegrapher, radio dispatcher, radio/telephone operator, news copywriter.
Sonar Technician	Operates electronic underwater detection and attack apparatus, obtains and interprets information for technical purposes, maintains sound detection equipment.	Normal hearing and clear speaking voice. Algebra, geometry, physics, electricity and shopwork desirable. Experience as amateur radio operator.	Oil well sounding device operator, radio operator, inspector of electronic assemblies, electronic technician, electrical repairer.

Career Fields	Duties & Responsibilities	Qualifications	Examples of Civilian Jobs
Storekeeper	Orders, receives, stores, inventories and issues clothing, foodstuffs, mechanical equipment and other items. In the Coast Guard also performs duties as disbursing clerk.	Typing, bookkeeping, accounting, commercial math, general business studies and English helpful.	Sales or shipping clerk, warehouse worker, buyer, invoice control clerk, purchasing agent.
Subsistence Specialist	Cooks and bakes, prepares menus, keeps cost accounts, assists in ordering provisions, inspects foodstuffs.	Experience or courses in food preparation, dietetics, and record keeping helpful.	Cook, pastry chef, steward, pie maker, butcher, chef, baker.
Telephone Technician	Installs, maintains and repairs all telephone, telegraph and teletype equipment, switchboards, public address systems and inter-office communications systems.	Aptitude for electrical and mechanical work, use of numbers in practical problems. Previous electrical experience helpful.	Electrician, electrical equipment inspector and many jobs which are in the civilian field of telephonic communications.
Yeoman	Clerical and secretarial, typing, filing, operating office and duplicating equipment, preparing and routing correspondence and reports, maintains records and official publications.	Same qualifications required of secretaries and typists in private industry; English, business subjects, stenography and typewriting helpful.	Office manager, secretary, general office clerk, administrative assistant.

Chapter VI

They Speak for Themselves, Past and Present

When you join the military, you join the ranks not only of your contemporaries but of men and women who have led patriotic lives in America's history. Certainly reading about the lives of the great can inspire us—especially in military service. And we emphasize that word *service.*

Some we will meet would emphasize the word *military*. Americans are in the military not because we like war but because we want peace and believe in defending our country. In fact, we believe in the human rights of all peoples, even though our first obligation is to America.

Generally Americans take pride in their young people in uniform and respect the wisdom and dedication of those who have served for many years. Just as in a family, coordination and cooperation are necessary, so if you join a military service, you owe it first your loyalty, then your energy and intelligence. After all, the making of a better world is not a task for one man or one woman but for all of us working together.

Minorities Make Military Careers

The military services offer significant opportunities to members of ethnic groups. For example, Earl Graves, editor and publisher of *Black Enterprise Magazine*, writes:

> Each month for the past ten years the pages of my magazine have recorded the successful ascent of black men and women up the corporate ladder and into the private world of business and finance.
>
> I know firsthand the skills and qualities that are necessary for success, and I also recognize that the experience offered by

> the Reserve Officers Training Corps are those that can guarantee you individual success and accomplishment. Not only does ROTC provide the opportunity for assisting you in your college education, it also assures you entry into the armed services as a second lieutenant with... satisfying leadership responsibilities.
>
> I know because I was part of the ROTC program in college, and I know for certain that the training I received there and in the Army as an officer played a part in preparing me for my career as a businessman and leader in the business community today.

If you are black and wondering about a military career, you might want to obtain the Army ROTC pamphlet "Double Your Chances for Success."

Here you could read comments from other black leaders such as Maurice Buchanan, Captain, U.S. Army, who writes:

> I turned down a basketball scholarship for an Army ROTC scholarship. Many friends have asked me why I made that decision. Well, I knew I needed to get that ticket punched to be successful. With Army ROTC I knew I'd have a job after graduation. And that's more than a lot of my peers could say.
>
> An Army officer's job is both challenging and rewarding. It's also unique in the amount of responsibility you're given coming right out of college.
>
> I may stay in the military. But if I decide to get out, I've got the best reference in the world—a commission in the United States Army. And I owe it all to Army ROTC.

Or from Leo A. Brooks, Brigadier General, U.S. Army, who affirms that motivation, self-discipline, and leadership are the qualities businesses look for and Army ROTC gives you.

> During my twenty-seven years in the Army, I have frequently reflected on the fond memories of those early ROTC days at Virginia State University. Perhaps the most common leadership quality that I recognized as the key to the success of those Army officers that I had met was confidence in themselves. I took that quality as a challenge for my own develop-

ment. I have found that those who imbue confidence in themselves earn the confidence of others.

Or Pamela M. Bland, First Lieutenant, U.S. Army:

Army ROTC was my choice for combining a college education with a military career. It also provided me with a foundation in leadership. Now, as an Army officer, I am exercising responsibility, making decisions and receiving invaluable administrative and management experience. These management skills, along with practical experience, are necessary for effective leadership and are easily transferable to top management positions in the private sector.

Or Hazel W. Johnson, Brigadier General, Army Nurse Corps:

The key to the Army's future is the professionalism of its officers. College curricula prepare students to be knowledgeable in their chosen field. Army ROTC provides the individual with that same knowledge and the skills to become a leader. The combination of university education and Army ROTC leadership training produces a well-rounded officer with all the qualifications required by today's officer corps. Our experience in the Army Nurse Corps has shown that professional nurses who have received their commissions through Army ROTC demonstrate the ability to perform as both officers and professional nurses immediately after starting active duty. Army ROTC provides newly commissioned Army Nurses careers ahead of their peers.

And from Raymond Woolfolk, Attorney, Eastern Airlines pilot:

Army ROTC graduates are "take-charge people." Upon graduation, the personal traits and abilities you carry forth into the world can open the doors to a wide range of career opportunities. The training I received in Army ROTC provided me with the basic leadership qualities, managerial skills and discipline necessary to meet the demanding challenges I face daily. In addition, Army ROTC greatly enhanced my ability to rapidly analyze courses of action, make sound decisions and quickly proceed with my mission.

And so it goes on—doctors, architects, superior officers, dentists, ministers tell their stories.

West Point Stories

Minority group members also played a part in America's military history in early days. Henry O. Flipper, born a slave in 1856 in Thomasville, Georgia, and educated by another slave at night in a woodshed, was the first black to graduate from West Point.

He entered the Academy in 1873, successfully completed the demanding four-year program, and was commissioned a second lieutenant in 1877. He wrote two books, one on his life at West Point and another about his experiences as an engineer in the Southwest. He served as assistant to the Secretary of the Interior and died in Atlanta, Georgia, at the age of eighty-four. Today he is remembered with an annual award presented to the cadet demonstrating exceptional qualities of leadership, self-discipline, and perseverance in the face of unusual difficulties.

And there was Charles Young who graduated in 1889 to begin an outstanding thirty-four-year military career. He became the first black in U.S. history to be appointed a military attaché when President Theodore Roosevelt sent him to Haiti in 1904. While there, he mapped much of the interior of Haiti, compiled a French-English-Creole dictionary, and wrote *The Military Morale of Nations and Races*, which was published in 1912.

While serving in the Philippines, he was appointed military attaché to Liberia, with the additional duty of training and reorganizing the Liberian Forces and Constabulary. Upon returning from Africa, he commanded one of the first Cavalry units to see action in Mexico. Brigadier General John J. Pershing listed Young among officers recommended to command a brigade with temporary promotion to general if the U.S. mobilized for World War I. Unfortunately Young became ill and was unable to accept that duty. He was promoted to colonel and retired from active duty in 1919. Returning to Liberia as a military attaché at the request of the Liberian government, he died three years later in Lagos, Nigeria, and was buried with full honors in Arlington National Cemetery.

And there was Vincent K. Brooks, a cadet captain, one of the first minority group members to hold the position of brigade commander while at West Point, who says:

> I can assure you that I was proud to be the first black cadet in West Point's long history to hold the position of Brigade

> Commander.... It is vitally important that we march on, continuing to turn society's obstacles into bridges, its walls into doors, its frustrations into opportunities, and its setbacks into challenges.... I recommend West Point, not for everyone, but for those who know they must walk an extra mile to receive the extra benefit of discipline and sound character.

David Moniac, born in 1802 in the Creek tribe, was appointed to the U.S. Military Academy under provisions of a 1791 treaty mandating the education of a limited number of Creek children at government expense. In Washington, D.C. he learned to read and write, and he entered the Academy a year later at age fifteen. The first American Indian to attend West Point, he became a major in the United States Army.

Patty Aceves, of Mexican heritage, tells us: "I've always tried my best and have never been shot down because I was a woman. I was brought up to think that the man is in charge of the family—but I was also taught that you can go as far as you want and as high as you want, even if you have to work hard to get there."

Patty has been proving it. She has mastered basic paratrooper skills; has served as an exchange cadet to a Mexican military college; has met the Mexican Secretary of Defense; and has spoken to a distinguished Mexican-American Women's National Association. In her spare time, she works with the Special Olympics, participates in the Big Brothers and Sisters program, and teaches Sunday School. She plans a career in Army medicine or a foreign field. She claims to have been shy in high school, but certainly West Point has developed her confidence.

Richard L. Garcia writes: "Every day at West Point is both challenging and rewarding. The standards of the Academy must be met each day, academically and physically... If you enjoy a challenging atmosphere and high standards, you may have what it takes to excel at West Point."

From Marine Corps History

During World War I, 305 women served briefly but efficiently in the Marine Corps. "Reservists (Female)," they were enrolled to do clerical jobs, freeing men for the front. The first of these forerunners of today's women Marines was Opha Mae Johnson, who enlisted in Washington, D.C., August 13, 1918, one day after the Honorable Josephus Daniels, then Secretary of the Navy, authorized the Navy and Marine Corps Reserves to accept women for service.

Recruiters were instructed to enlist only women of excellent character and neat appearance, with business and office experience. Stenographers, bookkeepers, and typists were in great demand. Most worked at Headquarters Marine Corps in Washington, but a few were stationed in recruiting offices as far away as San Francisco and Portland, Oregon.

By 1922, all of the women Reservists had packed away their uniforms and returned to civilian life. Many did accept Civil Service appointments at Headquarters Marine Corps, but from a military standpoint, for the next twenty years the Marine Corps remained strictly a man's world.

Then came World War II. The war in the Pacific was almost a year old before the Marine Corps again looked to woman-power to help out. On November 7, 1942, General Thomas Holcomb, Commandant of the Marine Corps, approved the formation of the United States Marine Corps Women's Reserve. The necessary legislation was sponsored by Congressman Melvin J. Maas of Minnesota, a former Marine Corps major general.

Mrs. Ruth Cheney Streeter of Morristown, New Jersey, was selected to head the program. The mother of three sons, all of whom saw war service, and a teenage daughter, Mrs. Streeter held a commercial pilot's license, was active in health and welfare activities, and was a member of a citizen's committee assisting servicemen stationed at Fort Dix. On January 29, 1943, she was commissioned a Major, USMCWR, and was sworn in as the first Director of the Women's Reserves. She served as Director throughout the war and achieved the grade of colonel.

Colonel Streeter, however, was not the first woman to go on active duty in the Marine Corps Women's Reserve. The first commission went to Anne A. Lentz, Women's Reserve representative for clothing. A civilian clothing expert who had helped the Army's WAC, Captain Lentz came to the Marine Corps in December 1942 on a thirty-day assignment to help design the uniform for the Women Reserve, and, as it turned out, was soon wearing the Marine Corps uniform herself.

Six other women received priority direct commissions to aid in setting up the new Corps. One of them, Captain Lillian O'Malley Daly, a former World War I Reservist, became MWR representative for West Coast activities.

Many women volunteered for Marine Corps duty and were trained by the Navy at Hunter College, New York, for enlisted women and Mount Holyoke College for officer candidates.

Seventy-five women (the first officer class) entered Mount Holyoke on March 13, 1943, and were commissioned on May 11. The first enlisted Marine class of 722 women entered Hunter College on March 26 and graduated on April 25, 1943.

By July 1943, the Marine Corps had its own training complex at Camp Lejeune, North Carolina, housing both the officer candidate and enlisted schools and Women's Reserve specialist schools.

Although there was skepticism at first, the women Marines became Marines in the true sense of the word, wearing the forest green and sharing the name Marines with the men.

Their recruiting slogan was "Free a Man to Fight." Within a year they were serving at every major post and station and at recruiting districts through the country. Besides the usual clerical jobs, they were assigned to such fields as communications, quartermaster, post exchange, motor transport, food services, personnel, intelligence, administration, recruiting, community relations, education, legal assistance, and photography. By June 1944, the women Marines made up 85 percent of the enlisted personnel on duty at Headquarters Marine Corps, and from half to two thirds of the personnel manning all major posts and stations.

By June 1944, women of the naval services (including the Marines) were permitted to serve on a volunteer basis anywhere in the Western Hemisphere including Alaska and Hawaii. By that time the Marine Women Reserves arrived in Hawaii January 1945—160 enlisted and five officers. Before the war ended, nearly a thousand women—two detachments—served with Marine Garrison Forces, Pearl Harbor, and at Marine Corps Air Station, Ewa, Oahu, Hawaii.

The Marine Corps Women's Reserve peak strength reached over 19,000—approximately the strength of a Marine Corps division. General Alexander A. Vandegrift, the second wartime Marine Commandant, remarked that the Women Reserves could "feel responsible for putting the 6th Marine Division in the field; for without the women filling jobs throughout the Marine Corps, there would not have been sufficient men available to form that division."

The second Director of the wartime Marine Corps Women's Reserve was Katherine A. Towle, who took office upon Colonel Streeter's resignation. She had joined as a captain early in 1943 to become Colonel Streeter's assistant. Colonel Towle, a Californian with bachelor's and master's degrees from the University of California, served from December 7, 1945, until the wartime office was phased out on June 14, 1946.

Major Julia E. Hamblet, who had only recently been released to inactive duty, was recalled to head the postwar Women's Reserve. On September 6, 1946, she was sworn in as Director. A 1937 graduate of Vassar College, she was among the original group commissioned in the Marine Corps Women's Reserve upon graduation from that first class at Mount Holyoke.

Public Law 625 was passed on June 12, 1948, authorizing acceptance of women into the Regular component of the Marine Corps. Basic ("boot") training for enlisted women, which had been deactivated at Camp Lejeune in 1945, was activated early in 1949 at the Marine Corps Recruit Depot, Parris Island, South Carolina. The first commanding officer there was Captain Margaret M. Henderson, who ten years later was appointed Director of Women Marines with the grade of colonel.

Colonel Henderson, now retired, has written: "I readily observed how having women train at a station where male recruits trained resulted in increased responsiveness by the women Marines to the requirements of the Marine Corps." She felt that "the acceptance, guidance and cooperation by the commanding general (then Major General Alfred H. Nobel) and his entire staff at Parris Island was the key to the success of the training program."

In June 1949 the Women Officers' Training Class was established at Marine Corps Schools, Quantico, with Captain (later Lieutenant Colonel) Elsie E. Hill as first commanding officer. They were received by Marine Corps General Lemuel C. Sheperd, who assured them of his support and added that there is "a definite place for women Marines during peace, as there was during war."

In April 1949 the first women Marine Reserve platoons were activated in Boston and Kansas City, and soon there were thirteen such units throughout the country as part of the men's organized Reserve. By August 1950, these thirteen platoons were called up to fill stateside billets during the Korean War.

By the time of the Vietnam War, approximately 2,700 women Marines were on active duty. Some served overseas. By 1970 some 200 women were serving in such places as England, the Dominican Republic, France, Germany, Belgium, Japan, and Hong Kong.

Throughout the Vietnam years, Colonel Barbara J. Bishop, and later Colonel Jeanette I. Sustad, served as Director of Women Marines.

Colonel Bishop served in that capacity from January 3, 1954, to January 31, 1969, reporting from duty in Naples, Italy. A graduate of Yale and the University of Chicago, she was a member of the second class at Mt. Holyoke in World War II.

She and three former Marine Corps Women Directors were present on November 8, 1967, when President Johnson signed Public Law 90-130 increasing, and more or less equalizing, promotion opportunities for women in the military. The President noted: "Our Armed Forces literally could not operate effectively or efficiently without our women... So both as President and as Commander in Chief, I am very pleased and very proud to have this measure sent to me by the Congress."

This law opened the way for the temporary appointment of women as Brigadier General or Rear Admiral.

The Vietnam War ended with the signing of the Paris Peace Accords on January 27, 1973. On February 1 Colonel Margaret A. Brewer became the seventh and final Director of Women Marines. A graduate of the University of Michigan, she had entered the Marine Corps during the Korean War. The post-Vietnam era was a time of change that brought many new opportunities for women. Now they could serve in all occupational fields except combat arms (which included infantry, artillery, armor, and pilot/air crew). They could command units other than women's units and could serve in specified rear-echelon elements of the Fleet Marine Force.

By June 30, 1977, as women were more fully integrated in the services, the offices of Directors of the Marine Corps and other women's services were disestablished. On May 11, 1978, Colonel Brewer was appointed to a general officer's billet as Director of Information with the rank of brigadier general, serving until her retirement in 1980.

In December 1980, President Carter signed into law the Defense Officers Personnel Management Act, which (among other personnel changes) provides an integrated promotion policy for men and women officers.

If a woman from colonial times—or even a pioneer servicewoman of World War I—could come back, she would be amazed at some of the occupations for women today: aircraft maintenance specialist, oceanographer, helicopter pilot, and many others, according to a recent Defense Department report.

It wasn't so long ago that we thought there never would be a woman brigadier general or admiral. Today more than 1,500,000 women serve on the defense team. Some 199,000 wear the uniforms of the Army, Navy, Air Force, and Marine Corps (and of course the Coast Guard). Some 340,000 are civilian employees of the Department of Defense, defense agencies, and military departments, and more than 1,000,000 are the wives of military personnel.

To those who might prefer us to return to the days of hoopskirts and fans in magnolia gardens, we might point out that in today's world some of our country's cleverest and most lethal enemies are women. It is only logical that the United States make use of woman-power to aid in preserving those traditions we hold most dear. We still like magnolias and moonlight, roses and romance, but we'll answer the call when needed.

If you could visit them in their offices or at their work stations, you might meet some of the following, women doing their job and doing it well:

Captain Nina L. Patterson, Pease Air Force Base Hospital, New Hampshire, obstetrics ward charge nurse.

Seabee Michaelea Bradley, Naval Air Station, Whidbey Island, Washington, who welds, cuts, and fits steel and hopes to become a Navy diver.

Airman 1st Class Kelly Greer, Beale Air Force Base, California, who troubleshoots satellite tracking systems used to detect submarine-launched ballistic missiles.

Skilled machinist Kim Hall, Army Rock Island Arsenal III, who works with computerized machine tools.

Spec. 5 Megan West Tierney, Army Fife and Drum Corps.

Vivian L. (Bibs) Reynard, wife of an Army colonel, elementary school teacher who has written a book on Army protocol.

Engineering draftsman Arline Steuwer, White Sands Missile Range, New Mexico, civilian and mother; though confined to a wheelchair since 1946, she works at a modified drafting table she designed to help her in her Defense Department job.

Dr. Brenda Swann Holmes, research chemist at the Naval Research Laboratory, Washington, DC.

Senior Airman Lisa A. Simerson, Alabama Air National Guard, the first person of her rank to be named Airman of the Quarter by the 117th Tactical Reconnaissance Wing, and the first woman employed as an aircraft welder in her unit.

Flora P. Paulino, Chief, Publishing Division, Headquarters, Aerospace Audiovisual Service, Norton Air Force Base, California, who provides administrative support for more than 1,500 personnel at fifty-eight locations around the world.

Peggy Blevins, Marine Air Station (Helicopter), Tustin, California, 1984 Spouse of the Year for being "Hostess of Dependents and improving quality of life for Marine family."

Donna M. Alvarado, Deputy Assistant Secretary of Defense for Equal Opportunity and Safety Policy.

Clair E. Freeman, Deputy Assistant Secretary of Defense for Civilian Personnel Policy and Requirements.

Brigadier General Wilma L. Vaught, USAF, Committee on Women in the NATO Forces.

Staff Sergeant Cynthia I. Shumway, Malmstrom Air Force Base, Montana, small arms training instructor, who was named 1983 Combat Arms Technician of the year by the 15th Air Force.

Chief Air Controller Joanne L. Slavin, Naval Air Station, Miramar, California, master training specialist, advanced from seaman to chief petty officer in eight years.

Spec. 5 Deloris Thomas, 352nd Maintenance Battalion, Macon, Georgia, at Perimeter defense lookout duty on annual Reserve training.

Sergeant Cheryl Stearns, six-time U.S. women's national parachute champion, first woman to join the Army's "Golden Knights" aerial demonstration team. During her two years on the team, she claimed more gold medals and world records than most chutists win in a lifetime.

Army Reserve Captain Wanda Jewel, Redstone Arsenal, Alabama, wife and mother, who developed a sniper training program for the 25th Infantry Division and won a bronze medal in the 1984 Olympics small-bore rifle competition.

Captain Millie Hughes-Fulford, Army Reservist, in training for the Army Spacelab mission.

Seaman Amy E. Kumler, deck department worker aboard the repair ship *USS Ajax*, San Diego; a Sailor of the Month, who plans a computer career.

Lieutenant Colleen Nevius, Naval Air Test Center, Patuxent River, Maryland. The first Navy woman to attend the Test Pilot School, she received her Navy wings in 1979. Lt. Nevius is the daughter of a retired Navy captain and the wife of a Navy lieutenant commander.

Lieutenant Beverly A. Abbott, *USS Ajax*, a qualified deep-sea diver since 1982 and also a qualified officer of the deck.

Marine Lance Corporal I. R. Akers, motor transport maintenance assigned to the Engineer Support Company, 8th Engineer Support Battalion, 2nd Force Service Support Group, Camp Lejeune, North Carolina.

Corporal Michele A. Yoho, administration chief, Maine Aircraft Group 11, 3d Marine Aircraft Wing, Marine Corps Air Station, El Toro, California.

Colonel Gail Reals, Assistant Chief of Staff, Manpower, Marine

Corps Development and Education Command, Quantico, Virginia, moved up from enlisted ranks to become one of the most senior women Marine Corps officers.

Captain Kathryn E. Sullivan, 172nd Tactical Airlift Group, Mississippi Air National Guard, nurse who in one year learned to fire handguns so well that she earned a position on the State Combat Pistol Team.

Sergeant Major Willella T. Williams, first woman to reach that rank in the Army National Guard.

Master Sergeant Gail E. Thompson, 1st sergeant of 108th Combat Support Squadron, New Jersey, first woman to reach that rank in the Air National Guard.

Three top-ranking officers of the Air Force

Brigadier General Mary A. Marsh, Director for Manpower and Personnel, Organization of the Joint Chiefs of Staff, Washington, DC.

Brigadier General Frances I. Mossman, Mobilization Assistant to Director for Programs and Evaluation, Headquarters, U.S. Air Force, Washington, DC.

Brigadier General Diann A. Hale, Chief, Air Force Nurse Corps, Office of the Surgeon General, U.S. Air Force, Washington, DC

Three high-ranking officers of the Army

Brigadier General Connie Lee Slewitzke, Chief, Army Nurse Corps.

Brigadier General Myrna Hennrich Williamson, Commanding General, Third Reserve Officer Training Corps Region, Fort Riley, Kansas.

Brigadier General Mildred Elaine Pons Hedberg, Director of Personnel and Inspector General, U.S. European Command.

Two high-ranking Navy women

Commodore Mary J. Nielubowicz, Director, Navy Nurse Corps, first woman to serve as deputy commander for health care operations, Naval Medical Command.

Commodore Grace M. Hopper, special technical adviser to the commander, Naval Data Automation Command, Developer of COBOL, a widely used computer language.

Chapter **VII**

Conclusions

"Think of three things—whence you came, where you are going, and to whom you must account."

—Benjamin Franklin

"As I would not be a *slave*, so I would not be a *master*. This expresses my idea of democracy. Whatever differs from this, to the extent of the difference is no democracy."

—Abraham Lincoln

"The most important motive for work in the school and in life is the pleasure in work, pleasure in its result and the knowledge of the value of the result to the community."

—Albert Einstein

"I don't know what your destiny will be, but one thing I know: the only ones among you who will be really happy are those who have sought and found how to serve."

—Albert Schweitzer

"Remember always that you have not only the right to be an individual, you have an obligation to be one. You cannot make any useful contribution in life unless you do this."

—Eleanor Roosevelt

"This above all: to thine own self be true. And it must follow, as the night the day, Thou canst not then be false to any man."

—William Shakespeare

Who will be the achievers of tomorrow? Did it ever occur to you that you might be one of them and that a Service career might be the gateway to your goals?

Have you ever had an inspiration to invent something? Have you ever watched a firefly and wondered what made its light? Scientists have, and have discovered secrets of nature's realm. The flight patterns of birds, even the herbs of the field, have all contributed to our knowledge. Have you ever watched an apple fall from a tree? A story told by his grandniece is that the great Isaac Newton first got the idea of gravity by watching an apple fall from a tree.

Be that as it may, we know that Newton was a noted English mathematical scholar, philosopher, physicist, and astronomer. His law of universal gravity was the first systemization of knowledge in the physical sciences. He introduced the basic principles of mechanics and also of the orderly system of the heavenly bodies and their rotations. In addition to watching apples fall, Newton must have watched the stars at night to come to the conclusion that the orderly orbits of planets were governed by mathematical law. Today we have much stronger telescopes, but modern achievements are built on the pioneering work of the past.

Did you ever watch the steam spouting from a teakettle and wonder if such power could be harnessed? The story goes that Robert Fulton did. An American engineer and inventor from Lancaster County, Pennsylvania, Fulton must have been a keen observer, for he first painted portraits and landscapes of the American countryside, then went to England to study with the American painter Benjamin West. In 1793 Fulton gave up painting to study mechanics and engineering. He was interested in canal navigation and better transportation systems. He received many patents: one for sawing marble and another for spinning flax and hemp. He invented a submarine and finally a boat propelled by steam. In 1807 the steamboat *Clermont* was launched on the Hudson river, and he received an American patent for his invention. Others had dreamed of using steam as power, but Fulton's was the first practical success. The *Fulton* was the first steam-propelled warship.

So busy have we been in this book alerting you to all the eligibility and enlistment requirements, the rules and regulations of military service that we may have neglected to let you know that it is not all toil and trouble, marching and cramming. If you go to West Point, you'll never be at a loss for social, cultural, and sporting activities. Touring companies present live drama, and famous lecturers are invited. You are near enough to New York City for trips to attend plays and concerts. Back at school there's the Glee Club; the West Point Band; facilities for golfing, hunting, sailing, water skiing, downhill and cross country skiing, football, and lacrosse; some ninety clubs including horseback riding, sky diving, rock combos, and aeronautics, and gospel choirs and church services.

At the Coast Guard Academy, spare time is limited in the first or "Swab" year, but sooner or later you will find time for the extracurricular and social functions: karate club, bowling, debate, radio club, bicycle club, hockey club, rugby team, Christian Fellowship, choirs, glee clubs, Windjammers Marching Band, Cadet Dance Orchestra, a magazine, and a yearbook.

We have already mentioned some of the social and sporting activities at the Air Force Academy, and everyone knows that the Navy and the Marine Corps are not to be left out when it comes to social clubs and cultural activities.

In military life, you'll meet as many different types of individuals as you would in civilian life. Of particular interest are the personal careers carved out from military careers. In *The Retired Officer* magazine of October 1984, Chuck Hartle wrote about the late Navy artist Arthur Beaumont, who is said to have "mastered the sea with watercolors." He worked from Antarctica to the tropics, and his paintings of ships and waves are treasured all over the world.

Band leaders and composers have also arisen from military ranks. The names of Glenn Miller and Col. Gabriel come up in the Air Force. There probably isn't a major symphony orchestra in America that doesn't have at least one member who was formerly with an Army band. Many former Army members hold music professorships at renowned colleges, universities, and conservatories.

If you're serious about an Army music career, you have to arrange for an audition through a recruiting officer or the bandmaster of an active Army band. You must be able to sight-read music, and you will be tested for other musical skills such as intonation, articulation, phrasing, interpretation, breathing, and range. After you've passed an audition and mental and physical tests, you enter the Army Band Program at the advanced pay grade of Private First Class (E-3). Current salary for that pay grade is $695.10 a month, and after performing you'll be eligible for an accelerated promotion to Specialist Fourth Class (E-4) at $738.00 a month. As an Army musician you perform on a professional instrument, the property of the Army. You wear the most up-to-date Army uniforms at no cost to you, and you have a chance to continue your education in music. The Army will pay up to 75 percent of your tuition for approved college, vocational, or technical courses taken off-duty. Travel may be a part of this career, as Army band members are stationed at thirty-six locations in the United States and thirteen overseas, including Germany, Hawaii, Panama, Japan, and Korea. Other benefits are room and board, medical and dental care, and a chance to accumulate money through savings and government benefits for even further education and more professionalism.

The 1st Earl of Lytton, who used the pen name Owen Meredith, wrote, "...civilized man cannot live without cooks." Nor can the military services live without cooks. The Navy has a rating called Mess Management Specialist, and it recently enacted changes to "improve the overall standards of service in enlisted and officer messes and quarters management through new MS training."

Yes, in spite of what you may have heard, military life is not all toil and trouble, marching and cramming. The services believe in taking care of their own. Dorothea Johnson is a Navy wife who has written a book and given courses to Waves (Navy women, now) in hostessing. Her names for chicken dishes, Breast of Chicken Madeira, Italiano, or Viennese, certainly lend an allure to that food and no doubt were gleaned from living all over the world as a Navy wife.

It may be reassuring to know that the military academies and the services in general are not looking for stars but a good balance of scholarship and athletics. Whatever opportunity you choose, you are the star of your own career, but you have to remember that everybody else is a star too. You may be dealing with computers, airplanes, typewriters, word processors, but in the final analysis your success will be conditioned in part on how well you get along with your co-workers, with your platoon leader, with your commanding officer.

How can that be done? Through cooperation, courtesy, willingness to accept a challenge and follow through whether in books or athletics. Abiding by the rules, being on time, in correct uniform, and following orders are the basics of preliminary military training.

The military doesn't really want zany characters such as TV likes to present as military types.

If you learn your lessons well, you will be able to hold your head high and be proud to represent your country and your service. And the rewards will come.

Even if you don't land in Annapolis or West Point, remember there is a place for you if you give it your best. Nothing less is worthy of you. A little job done well is better than a big job botched up.

Along the way we have included charts of pay scales and military insignia that may interest you. But remember, these things change. Once sailors were paid in grog, and once there were no airplanes for the military. What doesn't change is the need for character, for dedicated individuals, for real men—and now for real women.

EQUIVALENT RANKS OF COMMISSIONED OFFICERS

0-10	General	Admiral
0-9	Lieutenant General	Vice Admiral
0-8	Major General	Rear Admiral (upper half)

0-7	Brigadier General	Rear Admiral (lower half) and Commodore–1986 change pending
0-6	Colonel	Captain (Navy)
0-5	Lieutenant Colonel	Commander
0-4	Major	Lieutenant Commander
0-3	Captain	Lieutenant (Navy)
0-2	1st Lieutenant	Lieutenant (junior grade)
0-1	2nd Lieutenant	Ensign

MONTHLY BASIC PAY

YEARS OF SERVICE

PAY GRADE	UNDER 2	2	3	4	6	8	10	12	14	16	18	20	22	26
COMMISSIONED OFFICERS														
0-10	5069.40	5247.90	5247.90	5247.90	5247.90	5449.20	5449.70	5533.20	5533.20	5533.20	5533.20	5533.20	5533.20	5533.20
0-9	4493.10	4610.70	4708.80	4708.80	4708.80	4828.50	4828.50	5029.50	5029.50	5449.20	5449.20	5533.20	5533.20	5533.20
0-8	4069.50	4191.30	4290.90	4290.90	4290.90	4610.70	4610.70	4828.50	4828.50	5029.50	5247.90	5449.20	5533.20	5533.20
0-7	3381.60	3611.40	3611.40	3611.40	3773.10	3773.10	3992.10	3992.10	4191.30	4610.70	4927.50	4927.50	4927.50	4927.50
0-6	2506.20	2753.70	2934.00	2934.00	2934.00	2934.00	2934.00	2934.00	3033.60	3513.30	3693.00	3773.10	3992.10	4329.60
0-5	2004.60	2354.10	2516.40	2516.40	2516.40	2516.40	2592.90	2732.10	2915.10	3133.20	3313.20	3413.40	3532.50	3532.50
0-4	1689.60	2057.40	2194.80	2194.80	2235.30	2334.30	2493.30	2633.70	2753.70	2874.60	2954.10	2954.10	2954.10	2954.10
0-3	1570.20	1755.30	1876.50	2076.30	2175.60	2254.20	2375.70	2493.30	2554.80	2554.80	2554.80	2554.80	2554.80	2554.80
0-2	1369.20	1495.20	1796.10	1856.70	1895.70	1895.70	1895.70	1895.70	1895.70	1895.70	1895.70	1895.70	1895.70	1895.70
0-1	1188.60	1237.50	1495.20	1495.20	1495.20	1495.20	1495.20	1495.20	1495.20	1495.20	1495.20	1495.20	1495.20	1495.20
COMMISSIONED OFFICERS WITH MORE THAN 4 YEARS ACTIVE DUTY AS AN ENLISTED OR WARRANT OFFICER														
0-3E	0.	0.	0.	2076.30	2175.60	2254.20	2375.70	2493.30	2592.90	2592.90	2592.90	2592.90	2592.90	2592.90
0-2E	0.	0.	0.	1856.70	1895.70	1955.70	2057.40	2136.00	2194.80	2194.80	2194.80	2194.80	2194.80	2194.80
0-1E	0.	0.	0.	1495.20	1597.20	1656.00	1716.00	1775.70	1856.70	1856.70	1856.70	1856.70	1856.70	1856.70
WARRANT OFFICERS														
CWO-4	1599.60	1716.00	1716.00	1755.30	1835.10	1916.10	1996.50	2136.00	2235.30	2313.90	2375.70	2452.50	2534.70	2732.10
CWO-3	1453.80	1577.10	1577.10	1597.20	1616.10	1734.30	1835.10	1895.70	1955.70	2014.20	2076.30	2157.00	2235.30	2313.90
CWO-2	1273.50	1377.60	1377.60	1417.80	1495.20	1577.10	1636.80	1696.80	1755.30	1816.80	1876.50	1935.90	2014.20	2014.20
WO	1061.10	1216.50	1216.50	1317.90	1377.60	1436.70	1495.20	1557.30	1616.10	1675.80	1734.30	1796.10	1796.10	1796.10
ENLISTED MEMBERS														
E-9	0.	0.	0.	0.	0.	0.	1860.60	1902.90	1945.80	1990.50	2034.90	2074.50	2183.70	2395.80
E-8	0.	0.	0.	0.	0.	1560.60	1605.00	1647.00	1690.20	1734.60	1774.80	1818.30	1925.10	2139.90
E-7	1089.60	1176.00	1219.80	1262.40	1305.60	1347.00	1390.20	1433.40	1498.20	1540.80	1584.00	1604.70	1712.40	1925.10
E-6	937.20	1021.80	1064.40	1109.70	1150.80	1192.80	1236.60	1300.20	1341.00	1384.20	1405.20	1405.20	1405.20	1405.20
E-5	822.60	895.50	938.70	979.80	1044.00	1086.30	1129.80	1171.20	1192.80	1192.80	1192.80	1192.80	1192.80	1192.80
E-4	767.40	810.30	857.70	924.60	960.90	960.90	960.90	960.90	960.90	960.90	960.90	960.90	960.90	960.90
E-3	723.00	762.30	793.20	824.70	824.70	824.70	824.70	824.70	824.70	824.70	824.70	824.70	824.70	824.70
E-2	695.40	695.40	695.40	695.40	695.40	695.40	695.40	695.40	695.40	695.40	695.40	695.40	695.40	695.40
E-1	620.40	620.40	620.40	620.40	620.40	620.40	620.40	620.40	620.40	620.40	620.40	620.40	620.40	620.40

E-1 with less than 4 months — 573.60

NOTE: Pay is limited to $5533.20 by level V of the Executive Schedule

Monthly Basic Quarters Allowances

Pay Grade	Without Dependents Full Rate	Without Dependents Partial Rate	With Dependents
0-10	$537.30	$50.70	$660.90
0-9	537.30	50.70	660.90
0-8	537.30	50.70	660.90
0-7	537.30	50.70	660.90
0-6	493.20	39.60	599.40
0-5	465.30	33.00	552.30
0-4	426.60	26.70	504.90
0-3	345.30	22.20	420.90
0-2	278.10	17.70	360.90
0-1	238.50	13.20	323.70
W-4	391.20	25.20	453.90
W-3	330.30	20.70	405.90
W-2	297.00	15.90	379.50
W-1	251.40	13.80	330.90
E-9	315.30	18.60	429.90
E-8	292.20	15.30	400.50
E-7	249.30	12.00	372.60
E-6	221.40	9.90	337.80
E-5	204.90	8.70	300.30
E-4	177.60	8.10	259.50
E-3	172.50	7.80	238.50
E-2	146.40	7.20	238.50
E-1	133.50	6.90	238.50
E-1*	122.70	6.90	213.60

*Less than 4 months of service

Basic Allowance For Subsistence

Officers **$106.18/month**

Enlisteds:

	E-1 under 4 mos.	All others
When rations in kind are not available	$5.50	$5.72
When on leave or granted permission to mess separately	4.87	5.06
When assigned under emergency conditions where no government messing is available	7.28	7.57

Insignia of the United States Armed Forces

ENLISTED

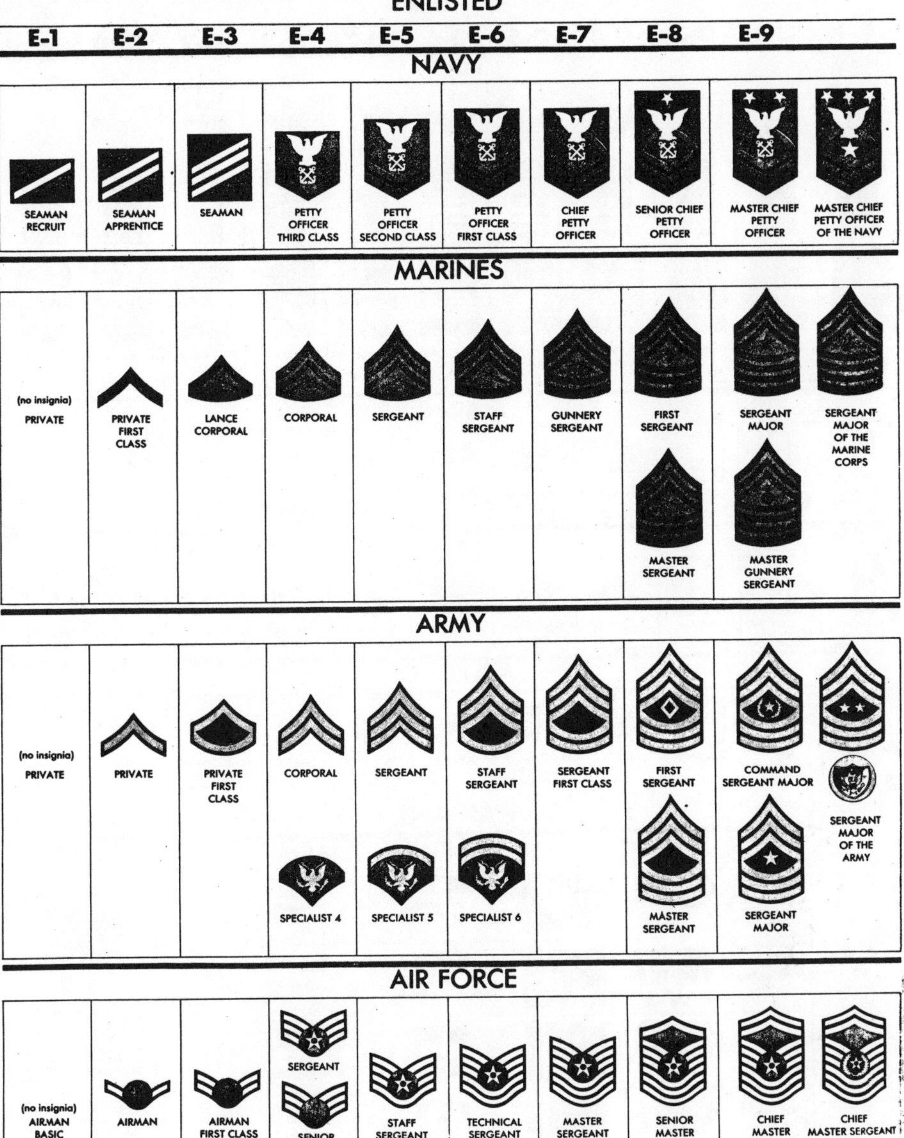

Chart by Phyllis Cox and John Pack

Insignia of the United States Armed Forces

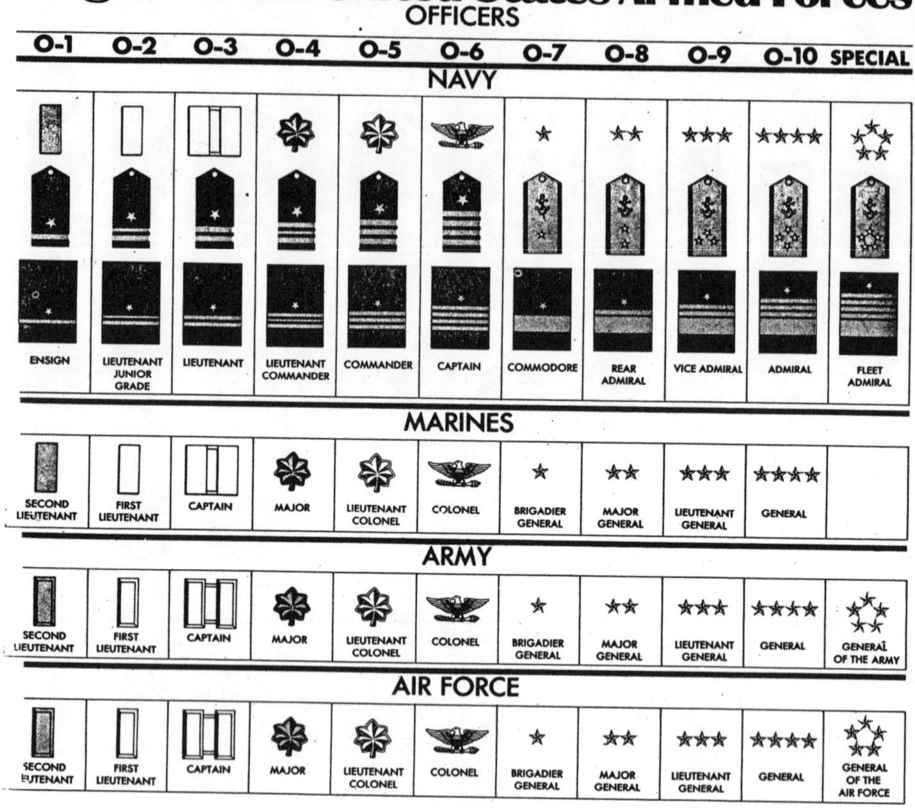

OFFICERS

	O-1	O-2	O-3	O-4	O-5	O-6	O-7	O-8	O-9	O-10	SPECIAL
NAVY	ENSIGN	LIEUTENANT JUNIOR GRADE	LIEUTENANT	LIEUTENANT COMMANDER	COMMANDER	CAPTAIN	COMMODORE	REAR ADMIRAL	VICE ADMIRAL	ADMIRAL	FLEET ADMIRAL
MARINES	SECOND LIEUTENANT	FIRST LIEUTENANT	CAPTAIN	MAJOR	LIEUTENANT COLONEL	COLONEL	BRIGADIER GENERAL	MAJOR GENERAL	LIEUTENANT GENERAL	GENERAL	
ARMY	SECOND LIEUTENANT	FIRST LIEUTENANT	CAPTAIN	MAJOR	LIEUTENANT COLONEL	COLONEL	BRIGADIER GENERAL	MAJOR GENERAL	LIEUTENANT GENERAL	GENERAL	GENERAL OF THE ARMY
AIR FORCE	SECOND LIEUTENANT	FIRST LIEUTENANT	CAPTAIN	MAJOR	LIEUTENANT COLONEL	COLONEL	BRIGADIER GENERAL	MAJOR GENERAL	LIEUTENANT GENERAL	GENERAL	GENERAL OF THE AIR FORCE

WARRANT

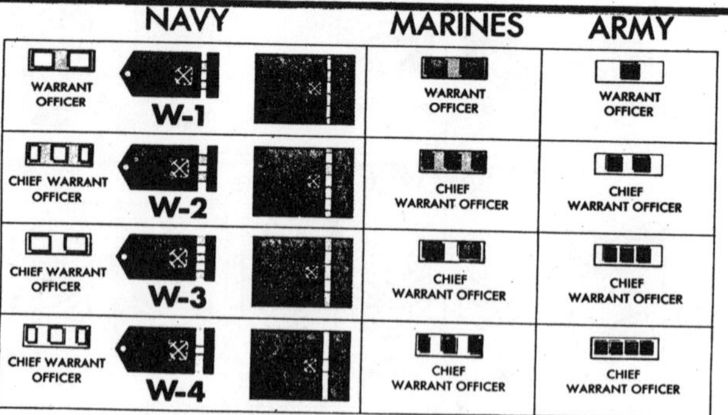

COAST GUARD

Coast Guard enlisted rating badges are the same as the Navy's for grades E-1 through E-6. E-7s through E-9s have silver specialty marks, eagles and stars, and gold chevrons. The badge of the Master Chief Petty Officer of the Coast Guard has a gold chevron and specialty mark, a silver eagle and gold stars. Coast Guard officers use the same rank insignia as Navy officers. For all ranks, the gold Coast Guard shield on the uniform sleeve replaces the Navy star.

Index